Engineering Descriptive Geometry

About the Author

Steve M. Slaby is Associate Professor of Graphics, Department of Civil and Geological Engineering, in the School of Engineering at Princeton University. He is a member of the American Society for Engineering Education, an Alumni Associate of the Institute of International Education, and an associate member of Sigma Xi.

Professor Slaby received the degree of Bachelor of Science in Mechanical Engineering from Lawrence Institute of Technology, and the degree of Master of Arts in Economics from Wayne University. He did postgraduate work in labor relations on a Fulbright fellowship at the University of Oslo in Norway.

In 1960 he had a Fellowship to do research in engineering graphics in Norway at the Norges Tekniske Høgskole. His work under this Fellowship also took him to the U.S.S.R. He has worked in industry as a member of several engineering departments, and has taught engineering drawing and descriptive geometry at Sampson College and at Lafayette College. Professor Slaby has had published several articles on Norwegian labor relations, and is coauthor of *Slide Rule Instruction Handbook*.

COLLEGE OUTLINE SERIES

ENGINEERING DESCRIPTIVE GEOMETRY

Steve M. Slaby
Associate Professor of Graphics
Princeton University

BARNES & NOBLE BOOKS
A DIVISION OF HARPER & ROW, PUBLISHERS
New York, Hagerstown, San Francisco, London

Formerly published under the title
DESCRIPTIVE GEOMETRY

Manufactured in the United States of America

85 86   10

To

Justine and Max

Preface

The science of descriptive geometry has been taught in engineering schools for many years. During these years it has evolved on both theoretical and practical levels to keep abreast with the ever-advancing science of engineering. The author's purpose in writing this book was to present a new approach to this rich subject. In this book an emphasis has been placed on presenting the basic principles of descriptive geometry in clear, annotated pictorial representations followed by orthographic examples. The orthographic examples include detailed explanations of each step required to develop a principle and to arrive at a solution of a problem.

In order that the pictorial representations attain a greater degree of three-dimensional effect, planes have been indicated as having thickness. In actual problem work, these planes are imagined as either being very thin or having no thickness.

It is hoped that this book will encourage the development of a graphic mind in the student. This is not a subject of techniques where mere memorization of procedures will give the student the graphic viewpoint he should attain. The student from the beginning must study this subject with one idea — the idea that everything he sees must be visualized in three-dimensional space. The student should try to develop this ability by analyzing and giving reasons for each step he takes in the solution of problems.

Problems to be solved by the student, at the end of each chapter, are set up on facsimiles of graph paper which can be easily reproduced. Many of these problems have answers given which will help the student to check his solutions. The Appendix contains examples of various applications of descriptive geometry principles to practical problems.

It is recommended that in addition to using this book the student refer to the excellent books listed in the Tabulated Bibliography, since different approaches to a subject are very helpful in gaining a clear understanding of the material.

I should like to thank my colleagues in the Graphics Department at Princeton University for their timely suggestions and their encouragement while this book was being written. Also I am very grateful to Dr. Gladys Walterhouse, Mathematics Editor of the publisher, for her cooperation and many suggestions while the manuscript was being prepared for publication.

STEVE M. SLABY

Table of Contents

Engineering Descriptive Geometry

I

The Concept of Descriptive Geometry

1. Definition of Descriptive Geometry. Descriptive geometry is the theory of engineering drawing. It is based on the principles of orthographic projection, which in turn are based on mutually perpendicular projection planes, and projectors which are perpendicular to these projection planes.

Descriptive geometry can also be defined as the graphical method of solving solid (or space) analytic geometry problems. By this definition the scope of this subject is greatly increased, and it becomes valuable in solving many engineering problems and a very useful tool in engineering research and development.

2. The Concept of Orthographic Projection — Definitions.

Lines of Sight. In orthographic projection the lines of sight of an observer, looking at a point or object in space, are assumed to be parallel and are represented by straight lines.

Projection Planes. Plane surfaces onto which a point or object in space is projected. These plane surfaces are imagined to exist between the observer and the point or object he is viewing. The observer's lines of sight are perpendicular to projection planes.

Projectors. Lines drawn from a point or object in space *perpendicular* to a projection plane.

Reference or Folding Line. The line of intersection between two *mutually perpendicular* projection planes. (Such lines are used as base lines from which all measurements are made in relation to the projections of a point or object to a projection plane.)

Plan or Top View (Horizontal Projection). A view in which the lines of sight of an observer are vertical. In this view a point or object in space is projected to a *horizontal* projection plane.

Elevation Views. Views in which the lines of sight of the observer are *level* or *horizontal*. The common elevation views are the front, left profile, right profile, and rear views. Any

1

elevation views which do not correspond to the common elevation views are called *auxiliary elevation views*. In these views a point or object in space is projected to auxiliary elevation projection planes that are *always perpendicular* to the *horizontal* projection plane. In *all* elevation views the *horizontal projection plane always appears as an edge*.

Inclined Auxiliary Views. Inclined views are views in which the lines of sight of an observer are neither vertical nor horizontal. In these views a point or object in space is projected to inclined projection planes which may be perpendicular to elevation projection planes or to other inclined projection planes.

3. Horizontal Projection Plane. Fig. 1 shows a point in space fixed relative to a projection plane. The projection plane in this figure can be considered similar to a ceiling in a room. The point, therefore, is located a specific distance below the ceiling. This distance is measured on a line (*projector*) which is drawn from the point perpendicular to the ceiling which is a horizontal plane.

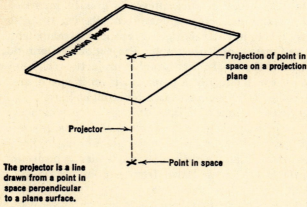

Fig. 1

4. Vertical Projection Plane. In order for the above point to be definitely fixed in space, its position relative to the walls of the room must be determined. In Fig. 2 it can be seen that the point in space is fixed relative to a horizontal plane and a vertical plane. The vertical plane can be considered similar to a wall in a room which is perpendicular to the ceiling. The distance from the point in space is measured on the perpendicular projector from the point to the wall.

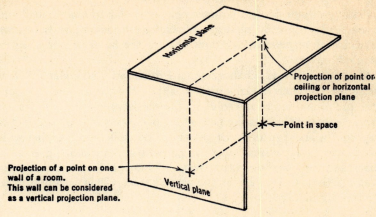

Fig. 2

5. Profile Projection Plane. The given point in space has been located relative to a horizontal projection plane and a vertical projection plane. This means that it cannot move up or down, or front or back, but it can move sideways. Therefore a third projection plane must be introduced so that the point will be fixed in all directions. This third plane can be compared with a wall in a room which is perpendicular to the ceiling (horizontal projection plane) and the first wall used (vertical projection plane). Fig. 3 shows this third projection plane, which is called a profile projection plane.

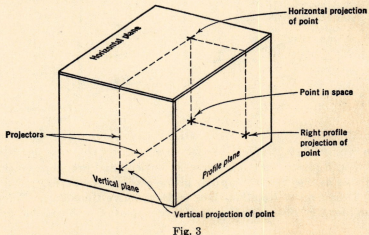

Fig. 3

The profile projection plane is perpendicular to the vertical and horizontal projection planes. *Mutually perpendicular projection planes and perpendicular projectors are the basis of orthographic projection.*

6. Application of Orthographic Projection Principles to Descriptive Geometry. In order to apply the principles of orthographic projection to the solving of descriptive geometry problems it is necessary that all the projection planes be represented in the plane of a sheet of drawing paper. How this is accomplished is best illustrated by imagining the projection planes in Fig. 3 to be hinged at the lines where they meet. These planes are then swung so that they will all lie in the same plane. This is represented pictorially in Fig. 4.

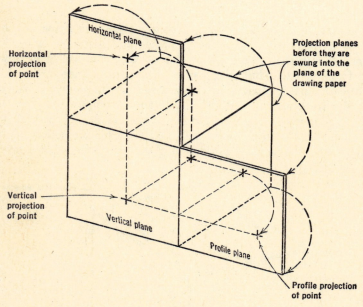

Fig. 4

Fig. 5 illustrates the horizontal, vertical, and profile projection planes as they actually appear in the plane of the paper.

7. System of Notation Used in Descriptive Geometry. Before problems can be solved using the principles of orthographic pro-

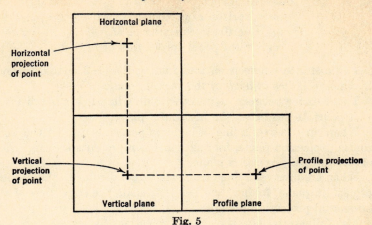

Fig. 5

jection, it is necessary that a methodical system of notation be adopted. Generally the following type of notation is used:

1. *Actual* points, lines, planes, or objects in space are denoted by *upper-case* letters.

2. *Projections* of points, lines, planes, or objects are denoted by *lower-case* letters with subindices indicating the projection planes in which they occur.

3. Projection planes are noted by numerals which are determined by the lines of intersection of the mutually perpendicular projection planes. These lines are referred to as *reference lines* or *folding lines* (see Fig. 6).

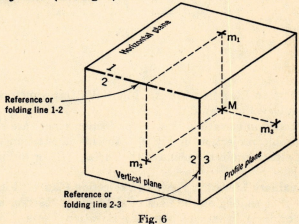

Fig. 6

m_1 = horizontal projection of M.
m_2 = vertical projection of M.
m_3 = right profile projection of M.

The distance between m_1 and folding line 1–2 is the distance that point M lies ***in back of*** the vertical plane.

The distance between m_2 and the folding line 1–2 is the distance that point M lies ***below*** the horizontal plane.

When an observer's line of sight is perpendicular to the horizontal projection plane, folding line 1–2 represents the edge view of the vertical projection plane.

When an observer's line of sight is perpendicular to the vertical projection plane, folding line 1–2 represents the edge view of the horizontal projection plane.

Fig. 7 illustrates how the actual projections (horizontal, vertical, and profile) of point M appear on the plane of a sheet of drawing paper. (Attention is called to the notation.)

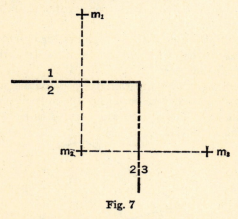

Fig. 7

8. Common Elevation Views. Fig. 8 illustrates pictorially the common elevation views.

The common elevation views or projection planes are shown in orthographic projection in Fig. 9. In this illustration the projection planes are in the plane of the paper. (Refer to Fig. 4 and Fig. 5.)

9. Auxiliary Elevation Views or Projection Planes. Figs. 8 and 9 show that as long as the lines of sight of an observer are horizontal or level, what are seen are elevation views. When an

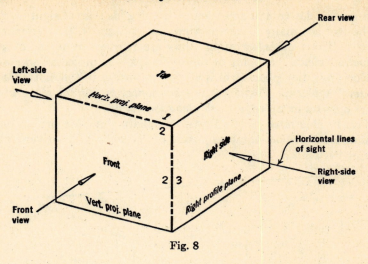

Fig. 8

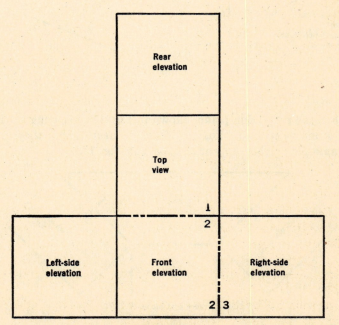

Fig. 9

observer circles around an object — keeping his lines of sight level — he will see many elevation views in addition to the common views. These other elevation views are referred to as ***auxiliary elevation views.*** A point or object in space is projected to elevation projection planes which are ***perpendicular*** to the ***horizontal projection plane.*** Therefore, the ***horizontal projection plane always appears as an edge in all elevation views.***

Fig. 10 illustrates pictorially two auxiliary elevation projection planes, a top or horizontal projection plane, and a front or vertical projection plane.

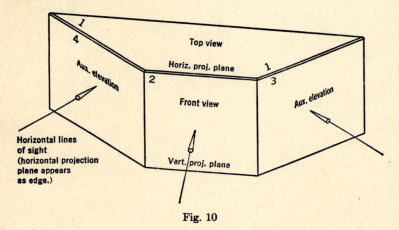

Fig. 10

When the projection planes shown in Fig. 10 are swung at their lines of intersection (folding or reference lines) so that they lie in the same plane, they appear as shown in Fig. 11.

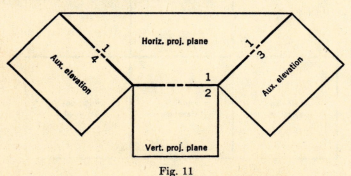

Fig. 11

In actual problem solution only the reference lines are used to indicate the different projection planes. (See Fig. 12.)

In Fig. 12 the projections of point *M* have been introduced. Note that the projectors from one view to another are **perpendicular** to the reference lines. Point *M* lies *X* distance below the horizontal projection plane. This distance is **always** seen in any elevation view, and in addition the horizontal projection plane always appears as an **edge** in all elevation views and is represented by reference lines.

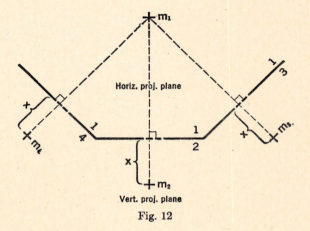

Fig. 12

If only the horizontal and vertical projections m_1 and m_2 are given and it is required to find an elevation view (3), the procedure is as follows:

1. Draw a projector from m_1 perpendicular to reference line 1–3.

2. Measure distance *X* below reference line 1–2 and transfer this distance to elevation view 3 by laying off this distance from reference line 1–3 on the projector from m_1.

10. Inclined Auxiliary Views or Projection Planes. Fig. 13 shows an inclined auxiliary projection plane which is perpendicular to the **vertical** projection plane. When an observer looks in a direction which is perpendicular to the inclined projection plane, the vertical projection plane appears as an edge to him. In this view he can see the distance that a point or object is located behind the vertical projection plane.

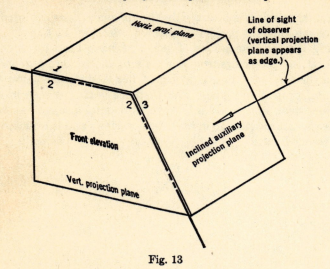

Fig. 13

Fig. 14 shows one inclined auxiliary projection plane which is perpendicular to the vertical projection plane, and the horizontal and vertical projection planes swung into the plane of a sheet of drawing paper. In this figure the projections of point M have been introduced.

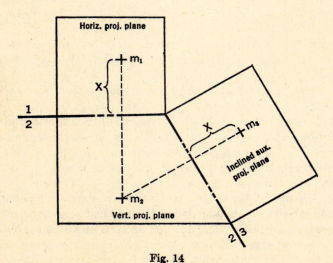

Fig. 14

Point *M* is *X* distance behind the vertical projection plane. This distance is seen whenever the *edge* of the ***vertical projection plane*** is seen. The vertical projection plane appears as an edge when an observer looks at the top (horizontal), bottom, left profile, and right profile projection planes in addition to inclined projection planes that are perpendicular to the vertical projection plane.

11. Key to the System of Orthographic Projection. The key to the system of projection used in descriptive geometry is pictorially illustrated in Fig. 15. It is suggested that the student thoroughly study the details of this figure, since once the principles that are involved here are digested, the remainder of this subject should be understood without too much difficulty.

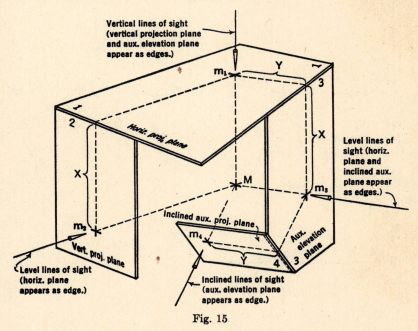

Fig. 15

Fig. 15 shows a series of consecutive projection planes where there are always two planes perpendicular to a third plane. Most of these projection planes are already familiar. For example, both the vertical projection plane and the auxiliary elevation plane are perpendicular to the horizontal projection plane. In addition the horizontal and inclined auxiliary projection planes (#1 and #4) are perpendicular to the auxiliary elevation projection plane (#3).

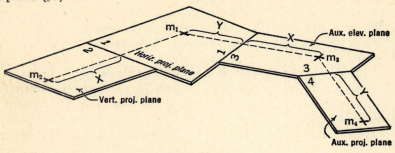

Fig. 16

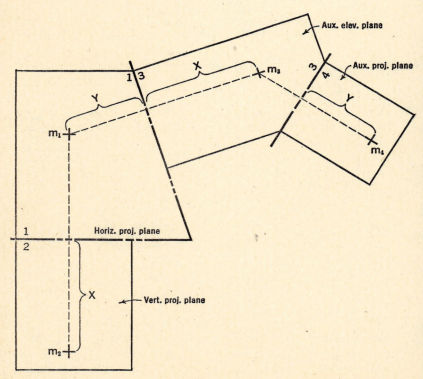

Fig. 17

Note that point M lies X distance below the horizontal projection plane. This distance is indicated on the vertical and the auxiliary elevation projection planes where the respective projections of point M appear (m_2 and m_3).

Also note that point M is Y distance behind the auxiliary elevation projection plane (#3). This distance appears on the horizontal and inclined auxiliary projection planes (#1 and #4) where the respective projections of point M appear (m_1 and m_4).

When the consecutive projection planes in Fig. 15 are swung into the same plane they appear as is pictorially illustrated in Fig. 16.

Fig. 17 shows the consecutive projection planes as they appear in the plane of a sheet of drawing paper. (In actual problem solution only the reference lines would be used to indicate the various planes.)

If the horizontal and vertical projections of point M (m_1 and m_2) are given and it is required to find the auxiliary elevation of this point as shown, the distance X is measured with dividers from the reference line 1–2 to the vertical projection m_2, and this distance is laid off on the auxiliary elevation plane from reference line 1–3 on the projector, and thus m_3 is located.

In other words, in order to find m_3 it is necessary to go back *two views* (horizontal and vertical) following the projectors to get the measurement which locates m_3.

This procedure applies to all problem solutions.

Apply this procedure to find how m_4 is located, keeping in mind the principles involved.

Practice Problems

Problems for solution are presented at the end of each chapter so that the student may test his knowledge of the material covered therein by solving as many of these problems as possible.

The problems are set up so that they can be reproduced and solved on $8\frac{1}{2}'' \times 11''$ cross-section paper having 4 squares to an inch. Whenever two sheets are required for a solution this is indicated in the problem statement. All measurements and location of points, lines, planes, etc., are to be made in full scale on the cross-section paper unless another scale is indicated in the problem statement. The problems should be set up and located on the paper exactly as shown so that the solution will remain on the given sheet. (Place paper on the drawing board with 11'' edge vertical.)

Answers are given to a representative number of problems. The accuracy of a graphical solution depends upon the accuracy with which a problem is reproduced, the quality of the instruments used, and the skill with which these instruments are handled. The student should allow for an error of at least plus or minus 1%.

In solving the following problems the student should make a habit of noting all given and constructed data in the field of the drawing. Construction lines should be light and sharp (use a 6H pencil); given lines, planes, etc. should be dark sharp lines (use a 2H pencil).

I-1. Refer to reference lines 1–2 and 2–3.
   (a) When viewing view #2 what does RL 1–2 represent?
   (b) When viewing view #2 what does RL 2–3 represent?
   (c) When viewing view #3 what does RL 2–3 represent?
   (d) When viewing view #1 what does RL 1–2 represent?

I-2. Given: Horizontal and vertical projections of point P.
   Find: Profile projection p_3.

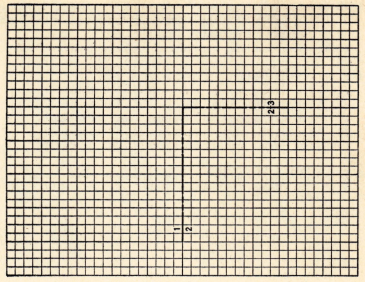

I–1

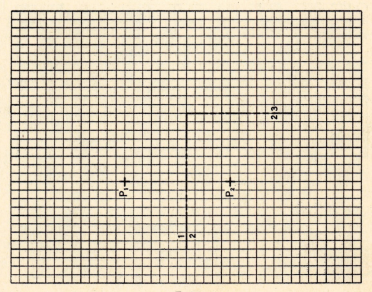

I–2

I-3. Given: The following data locate point Q: Point Q is located $1\frac{1}{2}$ inches below the horizontal projection plane and 1 inch behind the vertical projection plane.

Find: The horizontal, vertical, and profile projections of point Q.

I-4. Refer to reference lines 1–2, 1–3, 1–4, and 2–5.

(a) When viewing view #1 what does RL 1–3 represent?

(b) When viewing view #1 what does RL 1–4 represent?

(c) When viewing view #3 what does RL 1–3 represent?

(d) When viewing view #4 what does RL 1–4 represent?

(e) When viewing view #2 what does RL 2–5 represent?

I-5. Given: Horizontal and vertical projections of point P.

Find: Auxiliary elevation view (p_3) of point P.

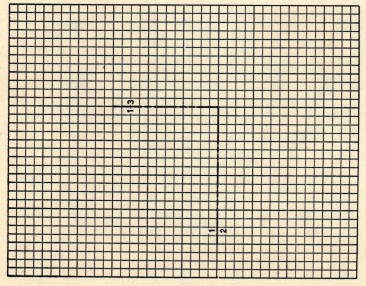

I–3

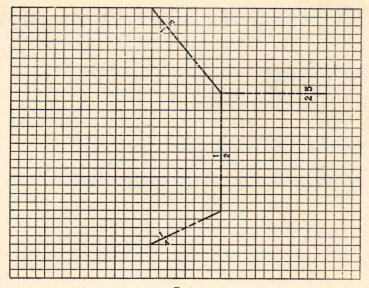

I–4

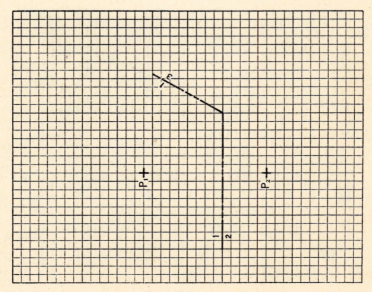

I–5

I-6. Given: Vertical and profile projections of point P.
     Find: Horizontal and auxiliary elevation projections p_1 and p_4 respectively.

I-7. Refer to reference lines 1–2, 2–3, and 2–4.
   (a) When viewing view #1 what does RL 1–2 represent?
   (b) When viewing view #2 what does RL 1–2 represent?
   (c) When viewing view #2 what does RL 2–3 represent?
   (d) When viewing view #2 what does RL 2–4 represent?
   (e) When viewing view #3 what does RL 2–3 represent?
   (f) When viewing view #4 what does RL 2–4 represent?

I-8. Given: Horizontal and vertical projections of point P.
     Find: Inclined projection p_3.

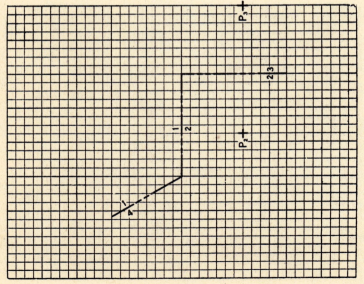

I–6

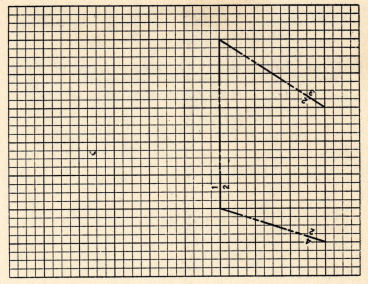

1–7

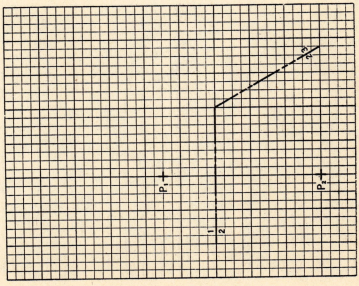

I–8

I-9. Given: Vertical and profile projections (p_2 and p_3) of point P.
  Find: Horizontal projection p_1, inclined projection p_4, and auxiliary elevation p_5 of point P.

I-10. Given: Horizontal and vertical projections (p_1 and p_2) of point P.
  Find: Projections of point P on auxiliary elevation plane #3, on auxiliary inclined plane #4, and on auxiliary inclined plane #5.

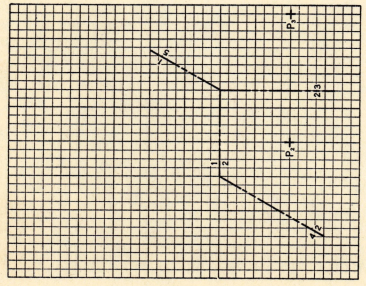

I–9

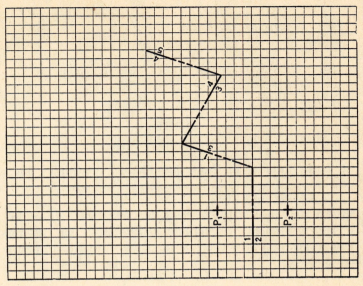

I–10

II

Lines

12. Lines Parallel to Projection Planes. A straight line will appear in true length when it is projected onto a projection plane which is *parallel* to the line.

13. Frontal Line. When a line is parallel to the *vertical* projection plane it is called a *frontal line,* and its projection will appear in true length (T.L.) on the vertical plane. See Fig. 18 (pictorial illustration).

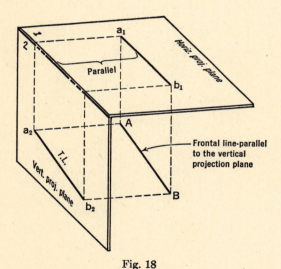

Fig. 18

Fig. 19 shows the frontal line in orthographic projection. The reference line 1–2 represents the edge view of the *vertical* projection plane when the line of sight is *perpendicular* to the horizontal projection plane.

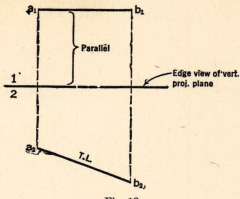

Fig. 19

14. Horizontal Line. When a line is parallel to the *horizontal* projection plane it is called a *horizontal line,* and its projection will appear in true length on the horizontal plane. See Fig. 20 (pictorial illustration).

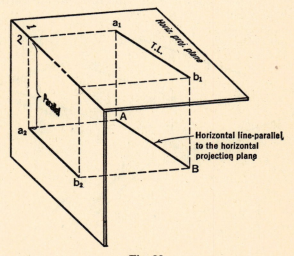

Fig. 20

Fig. 21 shows the horizontal line in orthographic projection. The reference line 1–2 represents the edge view of the *horizontal* projection plane when the line of sight is *perpendicular* to the vertical projection plane.

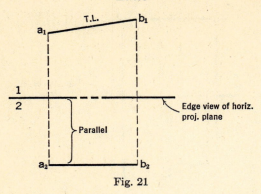

Fig. 21

15. Profile Line. When a line is parallel to a *profile* projection plane it is called a *profile line,* and its projection will appear in true length on a profile plane. See Fig. 22 (pictorial illustration).

Fig. 23 shows the profile line in orthographic projection. The reference line 2–3 represents the edge view of the *profile* projection plane when the line of sight is *perpendicular* to the vertical projection plane.

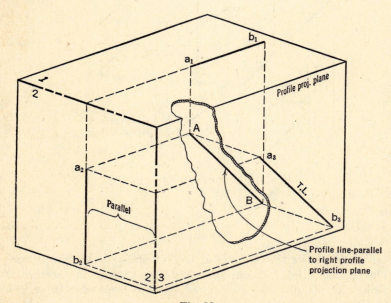

Fig. 22

Note that A lies X distance and B lies Y distance behind the vertical plane. These distances are also seen in the profile projection plane since the vertical projection plane appears as an edge when the line of sight is perpendicular to either the horizontal or profile projection plane.

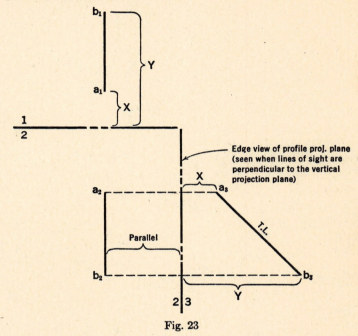

Fig. 23

16. True Length of a Line.
The projection of the true length of any line can be seen on a projection plane which is *parallel* to the line.

Example: True length of a line. (Refer to Fig. 24, p. 26.)

Given: Vertical and horizontal projections of line AB.

Find: The true length of AB.

Procedure:

    1. Pass a plane (represented by reference line 1–3) parallel to a_1b_1.

    2. Draw projectors from a_1b_1 perpendicular to reference line 1–3.

    3. From view #3 go back *two* views to get measurements to locate a_3b_3. (Refer to Fig. 17.)

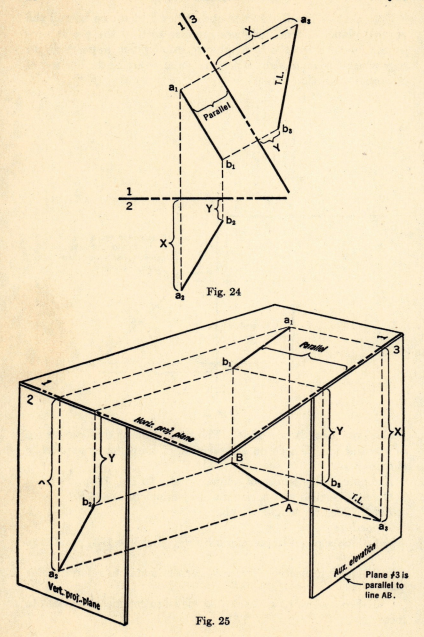

Fig. 24

Fig. 25

Plane ≠3 is
parallel to
line AB.

4. Transfer measurements (X and Y) to view #3, locating a_3 and b_3.

5. Connect a_3 and b_3 with a straight line and measure the true length.

See Fig. 25 for a pictorial representation of this solution.

The true length of line AB can also be found by passing a plane parallel to the vertical projection (a_2b_2) of the line. See Fig. 26.

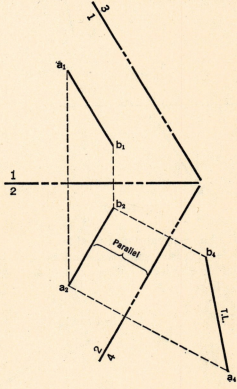

Fig. 26

17. True Slope of a Line. The slope of a line is defined as the angle a line makes with the *horizontal projection plane.* The slope is seen in a view where the line appears in *true length* and at the same time the *horizontal projection plane* appears as an *edge.*

The horizontal projection plane appears as an edge in elevation views; therefore the slope of a line can be seen when the *true length* of the line is found in an *elevation view*. See Fig. 27 and Fig 31.

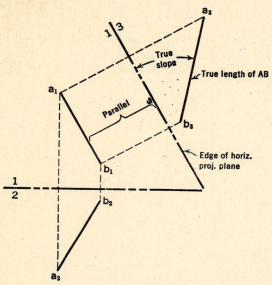

Fig. 27

The slope of a line is usually measured in degrees or is indicated as per cent grade. Per cent grade is the number of units of vertical rise for each hundred units of horizontal distance. (See Fig. 28.) Line *AB* has a 40 per cent grade.

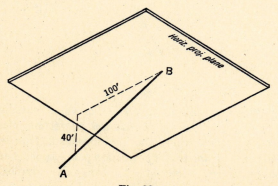

Fig. 28

18. Bearing of a Line.   The bearing of a line is the direction the *horizontal projection* of the line has relative to the points of a compass.

Fig. 29 shows line AB having a bearing of N 60° E. The horizontal projection (a_1b_1) of this line indicates that the line is headed north and 60° towards the east.

Example: Bearing, slope, and true length of a line.   (Refer to Fig. 30.)

Given: Horizontal and vertical projections of point A.

Find: The horizontal and vertical projections of line AB having a bearing of N 30° W, a slope of 45° down from A, and a length of $1\frac{1}{2}$ inches.

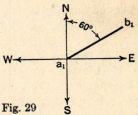

Fig. 29

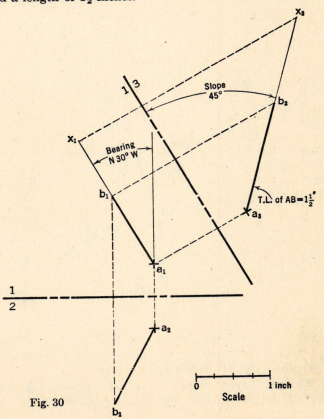

Fig. 30

Procedure:

1. From a_1 draw line a_1x_1 having a bearing of N 30° W and any *assumed* length.

2. Pass R.L. 1–3 parallel to a_1x_1 and locate a_3 by projection.

3. From a_3 draw a_3x_3 with a slope of 45° down from a_3. This is the T.L. of AX. (Slope is seen in this view since AX is T.L. and the horizontal projection plane appears as an edge.)

4. On a_3x_3 lay off $1\frac{1}{2}$ inches from a_3. This is the required line a_3b_3.

5. Project b_3 back to views #1 and #2.

19. Line as a Point. A line projects as a point on a plane which is *perpendicular* to the *true length* of the line. This is represented pictorially in Fig. 31.

Fig. 32 shows line AB as a point in orthographic projection.

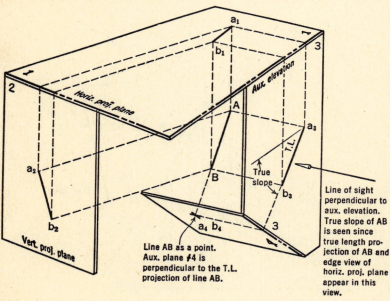

Fig. 31

20. Parallel Lines. Lines which are parallel in space will appear parallel in all views except in the views in which they appear as points or where one line is behind the other.

Fig. 33 shows parallel lines AB and CD in various projections to illustrate the general characteristics of parallel lines.

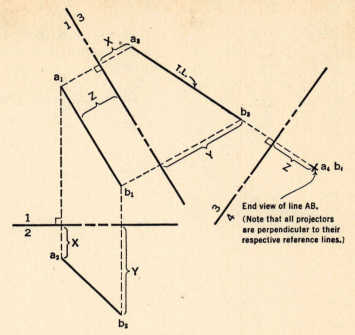

End view of line AB.
(Note that all projectors are perpendicular to their respective reference lines.)

Fig. 32

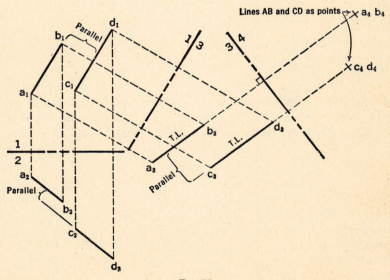

Lines AB and CD as points

Fig. 33

21. Perpendicular Lines. Lines which are perpendicular in space will appear perpendicular in any projection plane which shows at least *one* of the lines in *true length.* (An exception to this condition is when one of the lines appears as a point.)

Fig. 34 shows perpendicular lines AB and CD in space. These lines are projected onto a plane which is parallel to line AB, and therefore AB appears in true length on this plane.

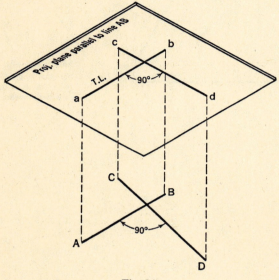

Fig. 34

Figs. 35 and 36 illustrate why lines which are perpendicular appear perpendicular when at least one of the lines is seen in true length. These figures and Fig. 37 are discussed on p. 34.

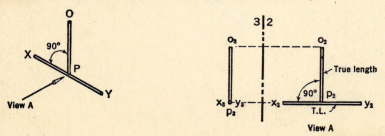

Fig. 35

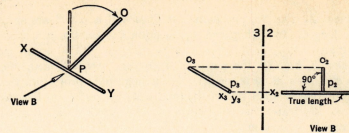

View B

Fig. 36

True length

View B

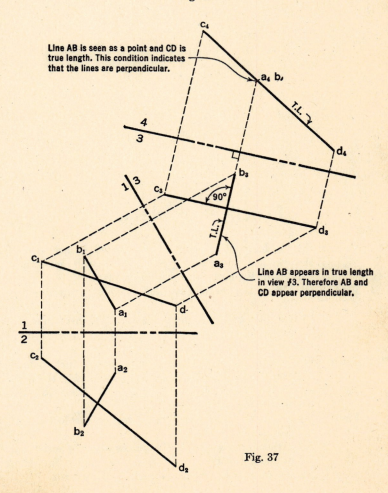

Line AB is seen as a point and CD is true length. This condition indicates that the lines are perpendicular.

Line AB appears in true length in view #3. Therefore AB and CD appear perpendicular.

Fig. 37

In view *A*, Fig. 35, perpendicular lines *XY* and *OP* are seen in true length. In orthographic projection the lines appear perpendicular.

In view *B*, Fig. 36, the position of *OP* has been changed, but *XY* is still seen as true length. In orthographic projection the angle remains 90° since one of the lines still appears in true length.

Fig. 37 illustrates the characteristics of perpendicular lines in orthographic projection.

Practice Problems

II-1. Given: Frontal line *AB* 2 inches long located $\frac{1}{2}$ inch behind the vertical projection plane and 1 inch below and parallel to the horizontal projection plane.

Find: (a) Draw the horizontal and vertical projections of line *AB*.

(b) Indicate by T.L. where the true length of line *AB* appears.

II-2. Given: Frontal line *CD* $2\frac{1}{2}$ inches long located 1 inch behind the vertical projection plane. *C* is $\frac{1}{2}$ inch below the horizontal projection plane and *D* is $1\frac{1}{4}$ inches below the horizontal projection plane.

Find: (a) Draw the horizontal and vertical projections of line *CD*.

(b) Measure the angle line *CD* makes with the horizontal projection plane.

(c) Indicate where the true length of *CD* appears.

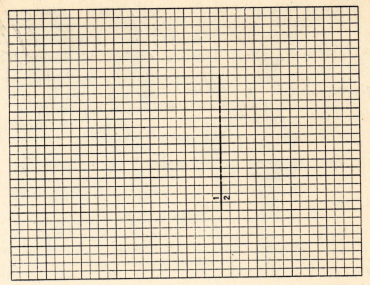

II–1

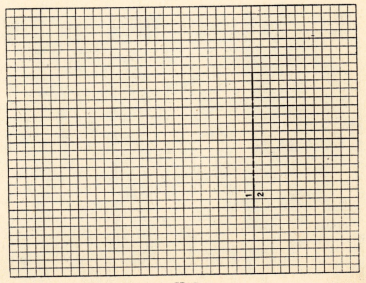

II–2

II–3. Given: Horizontal line *EF* 2½ inches long is located ½ inch behind and parallel to the vertical projection plane and 1 inch below the horizontal projection plane.

Find: (a) Draw the horizontal and vertical projections of line *EF*.

(b) Indicate where the true length of line *EF* appears.

II–4. Given: Horizontal line *FG* (2 inches long) is located so that it is 1¼ inches below the horizontal projection plane. *F* is located to the left of *G* and is ½ inch behind the vertical projection plane. *G* is located 1½ inches behind the vertical projection plane.

Find: (a) Draw the horizontal and vertical projections of line *FG*.

(b) Measure the angle line *FG* makes with the vertical projection plane.

(c) Indicate where the true length of *FG* appears.

II–5. Given: Profile line *AB* (2½ inches long) is located 1½ inches behind the profile projection plane. *A* is ¼ inch behind the vertical projection plane and ½ inch below the horizontal projection plane. *B* is 1¼ inches behind the vertical projection plane.

Find: (a) Draw the horizontal and vertical projections of line *AB*.

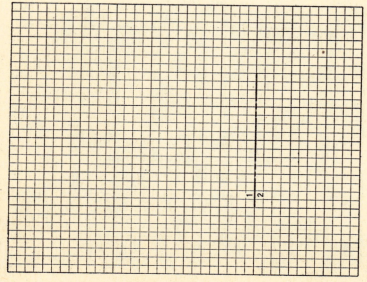

II–3

(b) Measure the angle line AB makes with the vertical projection plane.

(c) Indicate where the true length of line AB appears.

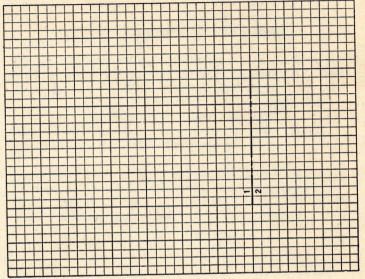

II–4

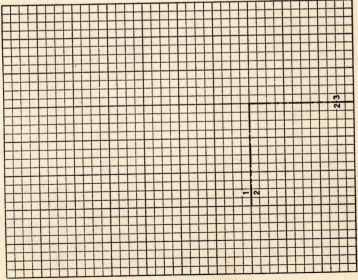

II–5

II–6. Given: Profile projection of profile line CD.
     Find: **(a)** Draw the horizontal and vertical projections of line CD.
           (b) What angle does CD make with the horizontal projection plane?
           (c) Measure and note the true length of CD.

II–7. Given: Horizontal and vertical projections of profile line EF.
     Find: **(a)** Describe in words the position of points E and F relative to the horizontal and vertical projection planes.
           (b) Draw the profile view of line EF.
           (c) Measure and note the true length of EF and the angle it makes with the vertical projection plane.

II–8. Given: Horizontal and vertical projections of profile line GH and the horizontal projection of point P on line GH.
     Find: The vertical projection of point P. (Suggestion: Draw the profile projection of line GH and point P.)

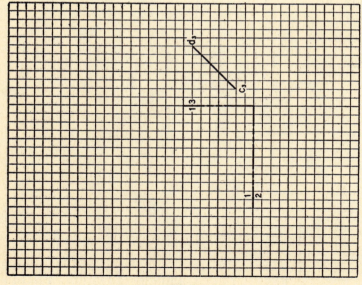

II–6

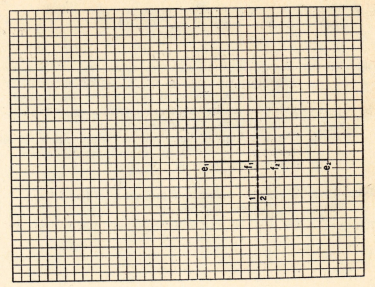

II–7

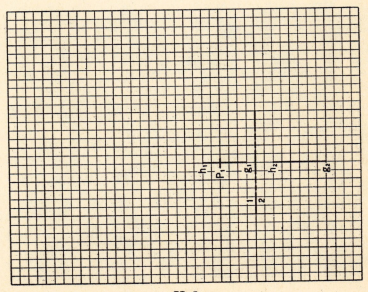

II–8

II–9. Given: Horizontal and vertical projections of line *AB*.
   Find: The true length of line *AB*. Measure and note this true length.

II–10. Given: Line *CD*. Point *C* is 1¼ inches behind the vertical projection plane and ½ inch below the horizontal projection plane. Point *D* is 1 inch to the right of *C*, ¼ inch behind the vertical projection plane, and 1½ inches below the horizontal projection plane.
   Find: The true length of line *CD* in an elevation view. Check this true length with an inclined view. Measure and note the true length in both cases.

II–11. Given: Horizontal and vertical projections of line *EF*.
   Find: (a) The true length of *EF*.
   (b) The true slope of line *EF* (indicate slope in degrees).

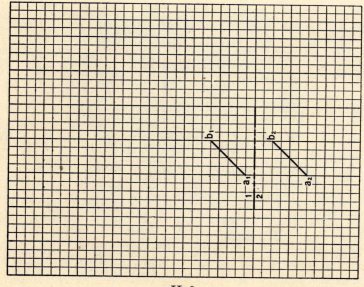

II–9

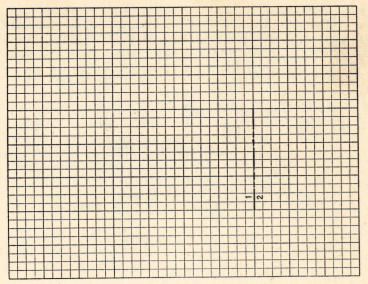

II–10

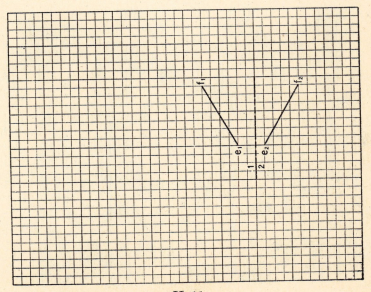

II–11

II-12. **Given:** Horizontal and vertical projections of point *G*.
   Find: (a) Draw the horizontal and vertical projections of a line *GH* which has a bearing of N 45° W, a slope of 30° downward from *G*, and a length of 1½ inches.
   (b) Indicate in degrees where the bearing and slope appear, and also indicate the true length of *GH*.

II-13. **Given:** Horizontal and vertical projections of point *A*.
   Find: (a) Draw the horizontal and vertical projections of line *AB* which has a bearing of N 30° E, a grade of 20% downward from *A*, and a length of 2 inches.
   (b) Indicate where the slope, bearing, and true length of line *AB* appear.
   (c) Locate the horizontal and vertical projections of point *P* on line *AB* which is 1 inch from *A*.

II-14. **Given:** Vertical and inclined projections of line *CD*.
   Find: The true length and slope of line *CD*. Measure and note the true length and slope on the drawing.

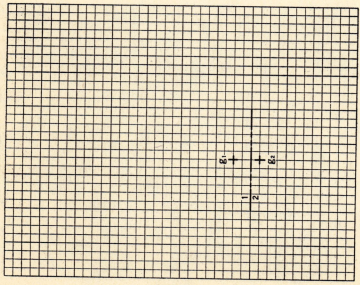

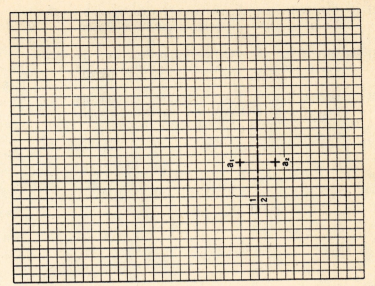

II–13

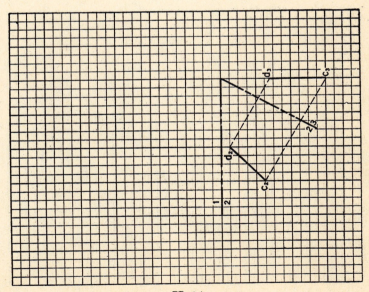

II–14

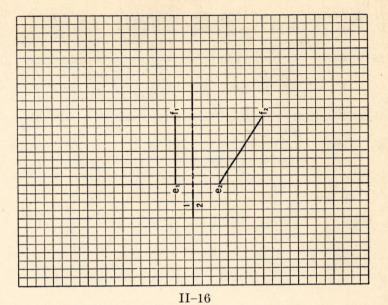

II–16

II–17

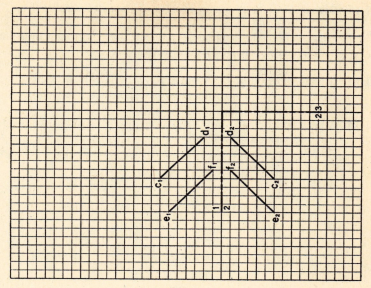

II–19

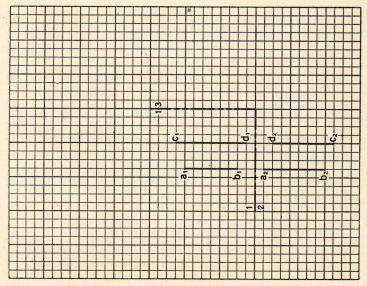

II–20

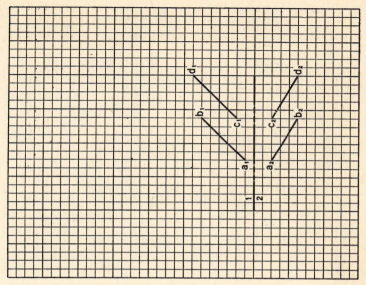

II–22

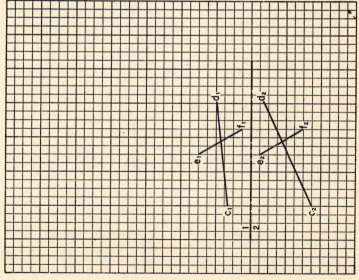

II–23

II–24. Given: Horizontal and vertical projections of two intersecting perpendicular lines *GH* and *JK*.

Find: Draw a view which shows both lines in true length.

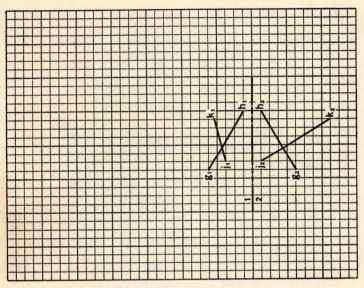

II–24

III

Planes

22. Definition of a Plane. A plane is a flat surface on which a straight edge or line will lie in any position.

23. Formation or Representation of Planes. Basically there are four ways in which planes can be represented:

(a) **By two intersecting lines.** Fig. 38 shows intersecting lines *AB* and *CD* forming plane *ACBD*. The plane is limited for drawing purposes but can be considered *limitless* depending upon the problem to be solved.

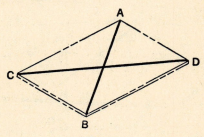

Fig. 38

(b) **By two parallel lines** (see Fig. 39). Parallel lines *EF* and *GH* form the plane *EFGH*.

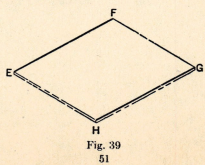

Fig. 39

(c) By three points not in a straight line (see Fig. 40). Points *X*, *Y*, and *Z* form the plane *XYZ*.

(d) By a point and a line (see Fig. 41). Point *X* and the line *AB* form plane *ABX*.

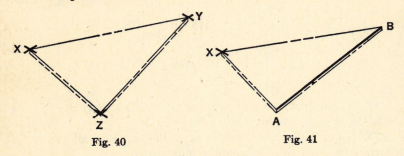

Fig. 40　　　　　　　　　　Fig. 41

24. Location of a Point in a Plane. If a point lies in a given plane any line may be drawn in the plane through the point. (See Fig. 42.)

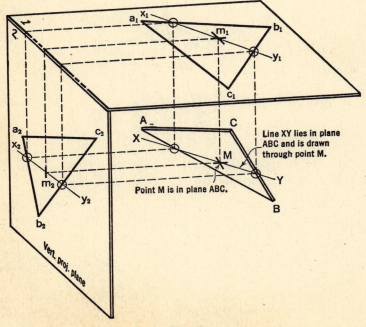

Line XY lies in plane ABC and is drawn through point M.

Point M is in plane ABC.

Vert. proj. plane

Fig. 42

Example: Location of a point in a plane. (Refer to Fig. 43.)
Given: Horizontal and vertical projections of the plane *ABC* and the horizontal projection (m_1) of the point *M*.
Find: The vertical projection of point *M* in the plane *ABC*.

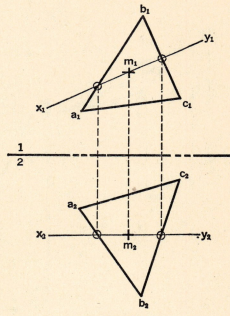

Fig. 43

Procedure:

1. Draw line x_1y_1 in plane $a_1b_1c_1$ and through m_1.

2. Find the vertical projection of line *XY* by dropping projectors from the points where x_1y_1 crosses a_1b_1 and b_1c_1 down to the front view (#2).

3. Draw a projector from m_1 to x_2y_2. Where this projector crosses x_2y_2 is the location of m_2, the required vertical projection of point *M*.

25. Edge View of a Plane. A plane can be seen as an edge in a view where *any* line which lies in the plane appears as a point. (See Fig. 44, p. 54.)

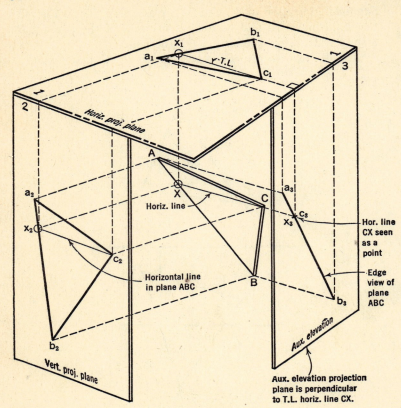

Fig. 44

Example: Edge view of a plane. (Refer to Fig. 45.)

Given: Vertical and horizontal projections of plane ABC.

Find: Edge view of plane ABC.

Procedure:

1. Draw horizontal and vertical projections of any line XY in plane ABC.

2. Find the true length of XY by passing a projection plane (1–3) *parallel* to x_1y_1.

3. Find XY as a point by projecting the T.L. of XY onto a plane (3–4) which is *perpendicular* to the T.L. (x_3y_3).

4. View #4 shows XY as a point; therefore plane ABC appears as an edge in this view.

By drawing a frontal or horizontal line in the plane ABC the edge view of this plane can be found more quickly and accurately since the true length of the line will appear in one of the given views. (See Fig. 44 and Fig. 46.)

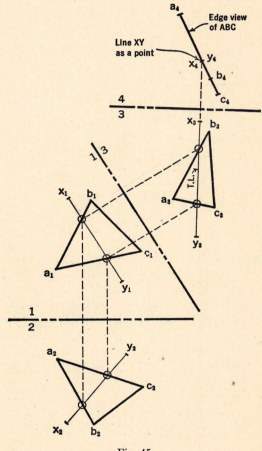

Fig. 45

Example: Edge view of a plane by seeing a horizontal line in the plane as a point. (Refer to Fig. 46, p. 56.)

Given: Vertical and horizontal projections of plane ABC.

Find: Edge view of plane ABC.

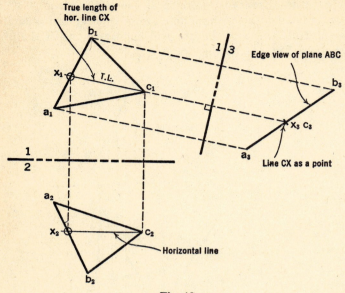

Fig. 46

Procedure:

1. Draw the horizontal and vertical projections of line CX in plane ABC (CX is taken as a horizontal line; therefore it appears as true length in the horizontal projection plane).

2. Pass a projection plane (1–3) perpendicular to the T.L. of line CX to see it as a point. In the point view of line CX (view #3) plane ABC appears as an edge.

26. True Slope or Dip of a Plane. The true slope or dip of a plane is the angle a plane makes with the *horizontal projection plane*. (See Fig. 47.)

The true slope of a given plane is seen in a view where the given plane appears as an *edge* and the *horizontal projection plane* also is seen as an *edge*.

Elevation views always show the horizontal projection plane as an edge; therefore the slope of a plane will always be seen in an elevation view.

Example: True slope or dip of a plane. (Refer to Fig. 48.)

Given: Vertical and horizontal projections of plane ABC.

Find: Slope of plane ABC.

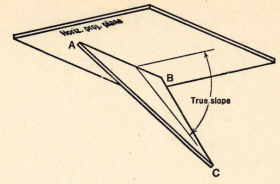

Fig. 47

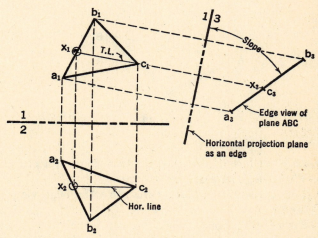

Fig. 48

Procedure:

1. Draw the horizontal and vertical projections of line CX in plane ABC (CX is taken as a horizontal line).

2. Find CX as a point so that plane ABC can be seen as an edge in an *elevation view*.

3. Measure angle (slope) that edge view of plane ABC makes with edge view of the horizontal projection plane.

27. True Shape or Normal View of a Plane. The true shape of a given plane is seen on a projection plane which is *parallel* to the given plane. (See Fig. 49, p. 58.)

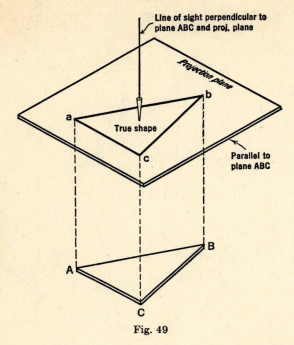

Fig. 49

General procedure used to find the true shape of a plane:

1. Find the given plane as an edge.

2. Pass a projection plane *parallel* to the edge view of the given plane.

3. Project the edge view of the given plane onto the parallel projection plane; the true shape appears on the parallel projection plane.

Example: True shape of a plane. (Refer to Fig. 50.)

Given: Vertical and horizontal projections of plane ABC.

Find: The true shape of plane ABC.

Procedure:

1. Draw horizontal and vertical projections of frontal line AX in plane ABC (a horizontal line could also be used).

2. Find AX as a point and therefore plane ABC as an edge.

3. Draw R.L. 3–4 parallel to the edge view of plane ABC.

4. Project edge view of plane ABC to projection plane #4. True shape of plane ABC is seen in this projection plane.

Note: All projectors are perpendicular to their respective reference lines.

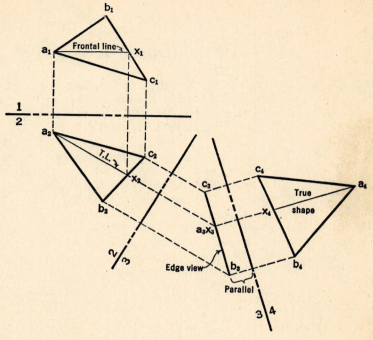

Fig. 50

Practice Problems

III-1. Given: Horizontal and vertical projections of parallel lines AB and CD and the horizontal projection of point P in the plane formed by lines AB and CD.

    Find: Draw the vertical projection of point P.

III-2. Given: Horizontal and vertical projections of three noncollinear points L, M, and N and the vertical projection of point P in the plane formed by points L, M, and N.

    Find: Draw the horizontal projection of point P.

III-3. Given: Horizontal and vertical projections of line EF and point X and the horizontal projection of point Q in the plane formed by EF and point X.

    Find: Draw the vertical projection of point Q.

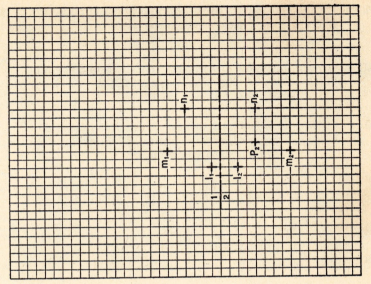

III–2

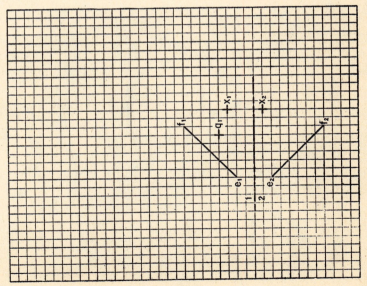

III–3

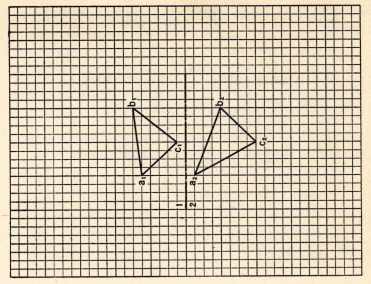

III–5

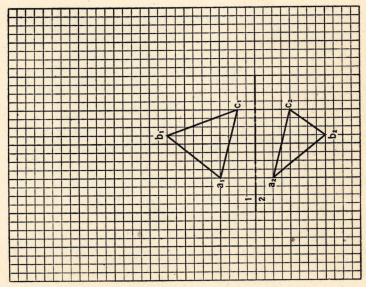

III–6

III–7. Given: Horizontal and vertical projections of plane *ABC*.
    Find: The slope of plane *ABC*. Measure and note the slope on the drawing.
    Answer: 54 degrees.

III–8. Given: Horizontal and vertical projections of plane *WXYZ*.
    Find: The true slope of plane *WXYZ*. Measure and note the slope in the drawing.
    Answer: 45 degrees.

III–9. Given: Horizontal and vertical projections of plane *ABC*.
    Find: (a) The slope of plane *ABC*.
          (b) The true shape of plane *ABC*.
    Answer: Slope = 40 degrees.

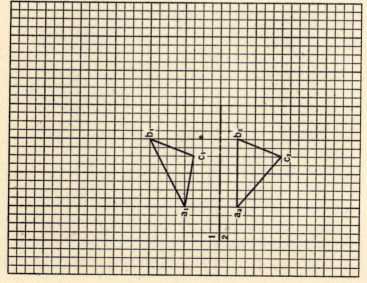

III–7

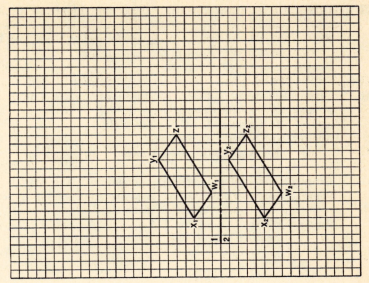

III–8

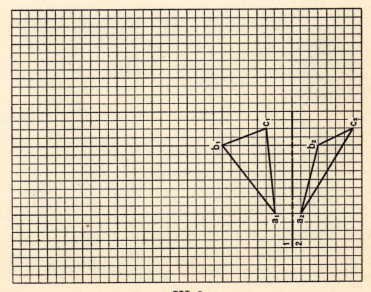

III–9

III-10. Given: Horizontal and vertical projections of points W, X, and Y.

Find: (a) The true slope of the plane formed by W, X, and Y.

(b) The true shape of the plane WXY.

Answer: Slope = 48.5 degrees.

III-11. Given: Horizontal and vertical projections of plane $ABCD$.

Find: (a) The true slope of plane $ABCD$.

(b) The angle ABC of plane $ABCD$.

Answer: (a) Slope = 35 degrees.

(b) Angle ABC = 88 degrees.

III-12. Given: Horizontal and vertical projections of plane ABC.

Find: The slope and bearing of a line drawn from A perpendicular to line BC in plane ABC.

Answer: Slope = 31 degrees.

Bearing = N 68° E.

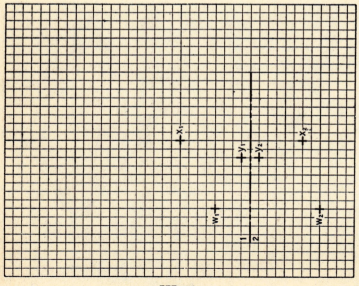

III-10

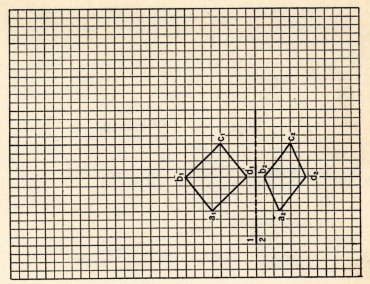

III–11

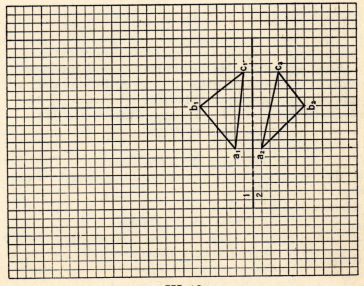

III–12

IV

Lines and Planes

28. Line Parallel to a Plane. If a line is parallel to any line in a plane it is parallel to the plane.   (See Fig. 51.)

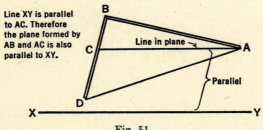

Line XY is parallel to AC. Therefore the plane formed by AB and AC is also parallel to XY.

Fig. 51

When the horizontal and vertical projections of two skew lines (lines that are *not* parallel and do *not* intersect) are given, it is possible to draw a plane containing one of the lines and parallel to the other by applying the principle illustrated in Fig. 51.

Example: Line parallel to a plane.   (Refer to Fig. 52.)

Given: Horizontal and vertical projections of two skew lines AB and XY.

Find: A plane containing line AB and parallel to XY.   Prove that the plane is parallel to the line by finding an edge view of the plane.

Procedure:

Creation of Parallel Plane
{
 1. Draw line AC through A on line AB and parallel to line XY in views #1 and #2.   (Two intersecting lines create a plane, and in this case the plane contains line AB and is parallel to line XY since one line in the plane is parallel to XY.)

Proof
of
Parallelism

2. Connect B and C with a straight line.

3. Draw a horizontal line and find its true length in view #1.

4. Pass a projection plane perpendicular to the true length horizontal line to see plane ABC as an edge (view #3).

5. In view #3 the edge view of plane ABC and line XY should appear parallel since plane ABC was originally created so that it contained AB and was parallel to XY.

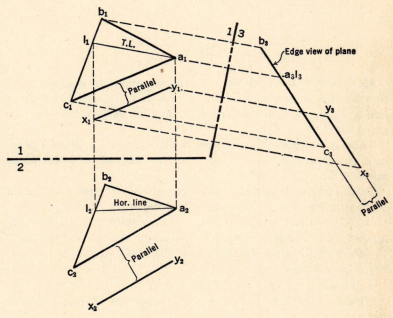

Fig. 52

29. Line Perpendicular to a Plane. If a line is perpendicular to a plane it is perpendicular to every line which lies in the plane at the foot of the perpendicular.

Fig. 53 shows line AB perpendicular to the given plane. All lines in the plane passing through B are perpendicular to line AB.

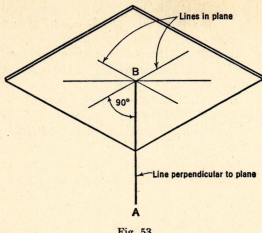

Fig. 53

30. Shortest Distance or Line from a Point to a Line. The shortest line that can be drawn from a point in space to a given line is a line which is drawn from the point *perpendicular* to the given line.

There are two methods by which the shortest line from a point to a given line can be found, the *line method* and the *plane method,* which are illustrated in Figs. 54 and 55.

In the line method the true length of the shortest line from a point to a line is seen in a view where the given line appears as a point. The plane method applies the principle that a line and a point form a plane. When the *true shape* of this plane is found, the shortest distance (or line) from the point to the line is seen in this true shape view.

Example: Line method. (Refer to Fig. 54.)

Given: Horizontal and vertical projections of line AB and point X.

Find: True length of the shortest line from point X to line AB. Show projections in all views.

Procedure:

1. Find T.L. of line AB. (Pass R.L. 1–3 parallel to a_1b_1.)

2. In view #3 draw a line from x_3 perpendicular to T.L. a_3b_3. This *locates* the shortest line, but the line does not appear in T.L. in this view. (See Fig. 36 in Chap. II.)

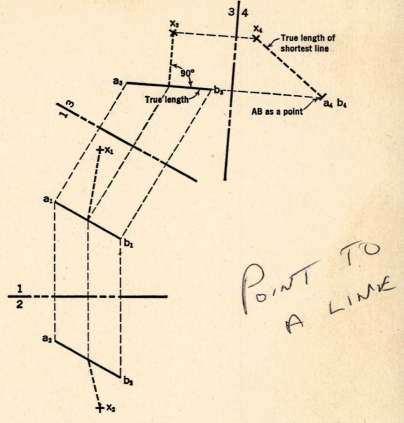

Fig. 54

3. Find line AB as a point. (Draw R.L. 3–4 perpendicular to T.L. a_3b_3.)

4. View #4 shows AB as a point, and the shortest line from X to AB appears in true length in this view.

Example: Plane method. (Refer to Fig. 55.)

Given: Horizontal and vertical projections of line AB and point X.

Find: The true length of the shortest line from point X to line AB.

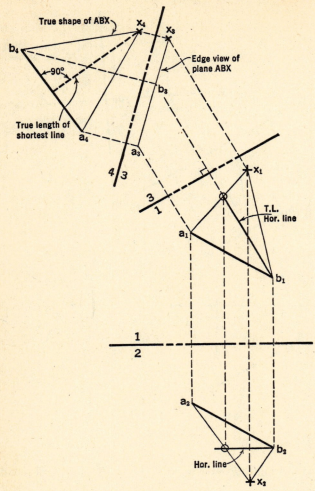

Fig. 55

Procedure:

1. Using line AB and point X create plane ABX.

2. Find edge view of plane ABX. (Draw horizontal line and see line as a point.)

3. Find true shape of plane ABX. (Pass R.L. 3–4 parallel to the edge view, $a_3b_3x_3$, of plane ABX.)

4. From x_4 in true shape view #4 draw a line perpendicular to a_4b_4. This is the required T.L. of the shortest line from point X to line AB.

31. Shortest Distance or Line between Two Skew Lines. The shortest line that can be drawn between two skew lines is a line which is *common* and *perpendicular* to both skew lines.

This common perpendicular has only *one* fixed position in space.

There are two methods by which the shortest line between two skew lines can be found, the *line method* and the *plane method,* which are illustrated in Figs. 56 and 57.

Example: Line method.   (Refer to Fig. 56.)

Given: Horizontal and vertical projections of two skew lines AB and CD.

Find: The shortest line between AB and CD.

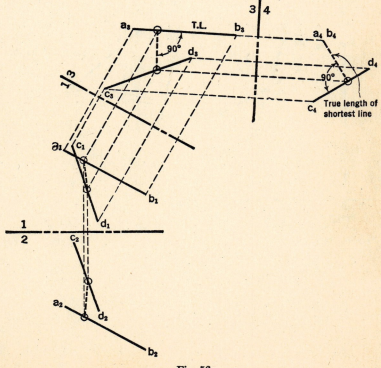

Fig. 56

Procedure:

1. Find the true length of one of the skew lines (true length of AB is used in example).

2. Show this true-length line (a_3b_3) as a point.

3. From a_4b_4 (point view) draw a line perpendicular to c_4d_4. This is the required line.

4. Project the shortest line back to all views.

Example: Plane method.   (Refer to Fig. 57.)

Given: Horizontal and vertical projections of two skew lines AB and CD.

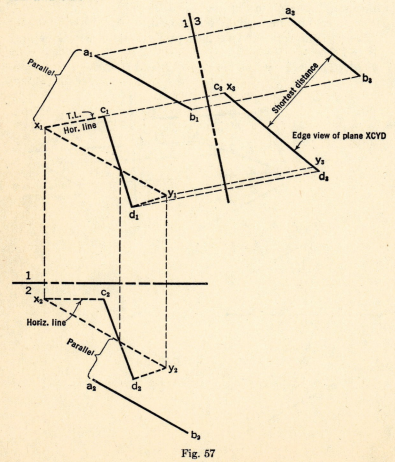

Fig. 57

Find: The shortest line between AB and CD.

If a plane is drawn containing one of the given **lines and parallel** to the other, an edge view of this plane will show the **actual shortest** distance between the two given lines.

Procedure:

1. Create a plane containing line CD and parallel to line AB by drawing any line XY through CD and parallel to AB (see Fig. 52).

2. Find edge view of this plane ($XCYD$) by first drawing horizontal line x_2c_2 and finding the T.L. of this horizontal line. Then see this line as a point (x_3c_3) in view #3.

3. Lines c_3d_3 and a_3b_3 appear parallel in view #3. The shortest distance is measured here.

32. Shortest Horizontal Line between Two Skew Lines. The true length of the shortest horizontal line between two skew lines is first seen in an *elevation view* where the two skew lines *appear* to be parallel.

The true length of this line will also appear in the horizontal projection plane (see Figs. 58 and 59).

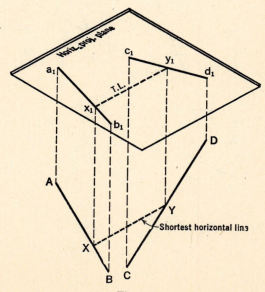

Fig. 58

Example: Shortest horizontal line between two skew lines. (Refer to Fig. 59.)

Given: Horizontal and vertical projections of two skew lines AB and CD.

Find: The shortest horizontal line between AB and CD.

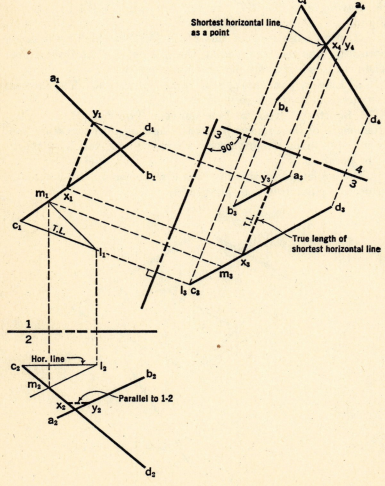

Fig. 59

Procedure:

1. Create a plane (CLM) containing CD and parallel to AB.

2. Find the edge view of this plane. In view #3 lines a_3b_3 and c_3d_3 appear parallel.

3. To **locate** the shortest horizontal line draw a projection plane (#4) perpendicular to view #3.

4. In view #4 lines a_4b_4 and c_4d_4 appear to intersect. At this **apparent** intersection the shortest horizontal line appears as a point (x_4y_4).

5. Project the shortest horizontal line back to the other views.

33. Shortest Grade Line between Two Skew Lines. The shortest grade line between two skew lines is a line which has a specified slope and intersects both skew lines at **one** particular location.

The true length of the shortest grade line is seen in a view where the two skew lines **appear** parallel. (See Fig. 60.)

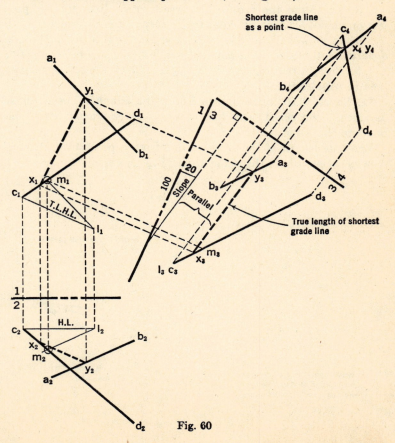

Fig. 60

The method of finding the true length of the shortest grade line is similar to that used in finding the shortest horizontal distance between two skew lines.

Example: Shortest grade line between two skew lines. (Refer to Fig. 60.)

Given: Horizontal and vertical projections of two skew lines AB and CD.

Find: The shortest grade line having a downward slope of 20 per cent between the two skew lines.

Procedure:

    1. Create a plane (CLM) containing CD and parallel to AB.

    2. Find edge view of this plane. In view #3 a_3b_3 and c_3d_3 appear parallel.

    3. To *locate* the required shortest grade line draw a projection plane (#4) perpendicular to the *specified slope* (see view #3).

    4. In view #4 lines a_4b_4 and c_4d_4 appear to intersect. At this *apparent* intersection the required grade line appears as a *point.*

    5. Project required grade line back to view #3, where it is seen in true length.

34. Shortest Distance from a Point to a Plane. The shortest distance is measured on a line which is *perpendicular* from the point to the plane (see Fig. 61).

The true length of the shortest line from a point to a plane is seen in a view where the plane appears as an *edge.*

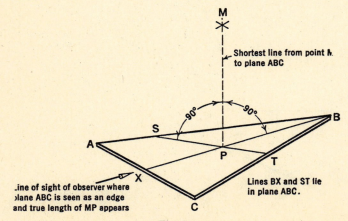

Fig. 61

If a line is perpendicular to *two intersecting* lines which *lie in a plane* it is perpendicular to the plane (refer to Fig. 61 and also see Fig. 53).

Example: Shortest line from a point to a plane.   (Refer to Fig. 62.)

Given: The horizontal and vertical projections of plane *ABC* and point *M*.

Find: The true length of the shortest line from point *M* to plane *ABC*.

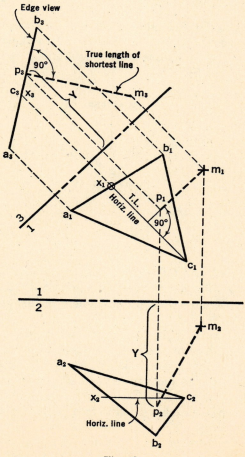

Fig. 62

Procedure:

1. Find edge view of plane ABC (horizontal line seen as a point is used in example).

2. From m_3 in view #3 draw a perpendicular line to $a_3b_3c_3$ (edge view of plane ABC). This line intersects the edge view at point p_3.

3. Draw m_1p_1 perpendicular to the true length horizontal line c_1x_1 (see Fig. 61). Locate p_1 by projecting back from view #3.

4. Locate p_2 in vertical projection plane by projecting down from the horizontal projection plane and measuring distance Y in view #3.

5. Measure the shortest distance m_3p_3.

35. Determination of Visibility between Two Skew Lines. The visibility of two skew lines is usually determined by inspection or observation. (See Fig. 63.)

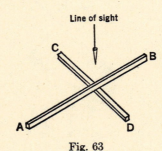

Fig. 63

In Fig. 63, AB and CD are two wooden sticks which represent two skew lines. From the line of sight indicated it is obvious that line AB is above CD. This is a case where visibility is determined by inspection.

In cases where visibility cannot be determined by inspection it can be determined by the method illustrated in Fig. 64.

Example: Visibility of two skew lines. (Refer to Fig. 64.)

Given: Horizontal and vertical projections of two skew lines AB and CD.

Find: Which line is above and in front of the other.

Procedure:

1. At the **apparent** point of intersection of AB and CD in the top view (#1) assume point x_1 **on** a_1b_1 and point y_1 **on** c_1d_1.

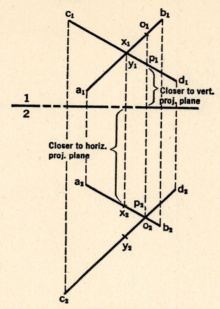

Fig. 64

2. Project x_1 and y_1 to their respective lines in view #2. It can be seen that x_2 is above y_2; from this it follows that AB is above CD.

3. At the **apparent** point of intersection of AB and CD in the front view (#2) assume o_2 on $a_2 b_2$ and p_2 on $c_2 d_2$.

4. Project o_2 and p_2 to their respective lines in view #1. It can be seen that p_1 is closer to the vertical projection plane than o_1. Therefore CD is in front of AB.

In determining the visibility of a solid the procedure described in Fig. 64 can be applied.

Example: Visibility of a solid. (Refer to Fig. 65 and Fig. 66.)
Given: Horizontal and vertical projections of a cube.
Find: Visibility of the cube.
Procedure: See procedure applied to Fig. 64.

Fig. 65 shows the horizontal and vertical projections of a cube with points X, Y, W, and Z assumed at the apparent intersections of the respective edges of the cube.

Fig. 66 shows the completed visibility of the cube in both views. The outside or boundary lines of the cube are visible in each view.

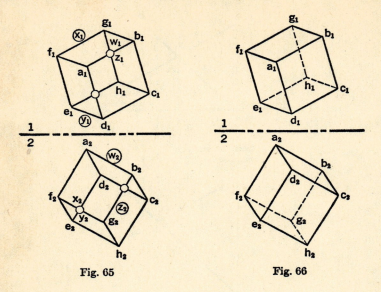

Fig. 65   Fig. 66

36. Piercing Point or Point of Intersection of a Line and a Plane.

When a straight line intersects a plane it pierces the plane at *one* point.   (See Fig. 67.)

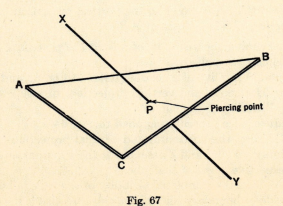

Fig. 67

In orthographic projection there are two main methods by which the piercing point of a line and a plane can be found: the *vertical cutting plane method* and the *edge view method*. These methods are illustrated in Figs. 68, 69, and 70.

A. Vertical Cutting Plane Method. Refer to Fig. 68.

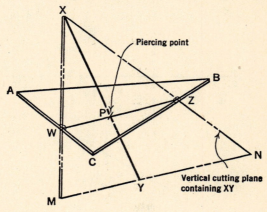

Fig. 68

Line XY intersects plane ABC at point P. If a *vertical* cutting plane MNX is passed through line XY it will intersect plane ABC on the line WZ. The piercing point P will lie on line XY **and** on the line of intersection WZ.

Fig. 69 shows how the vertical cutting plane method is used to find the piercing point in orthographic projection.

Example: Vertical cutting plane method. (Refer to Fig. 69.)

Given: Horizontal and vertical projections of plane ABC and intersecting line XY.

Find: The point (P) of intersection of the line and the plane.

Procedure:

1. Pass a vertical cutting plane (m_1n_1) through line x_1y_1 and intersecting plane $a_1b_1c_1$. The vertical cutting plane appears as a straight line in the horizontal projection plane.

2. The line of intersection of the plane ABC and the cutting plane MN is w_1z_1 in the horizontal projection plane. Locate this line of intersection in the vertical projection plane by projection.

3. Where w_2z_2 intersects line x_2y_2 is the location of the piercing point p_2 since this point is common to XY and WZ.

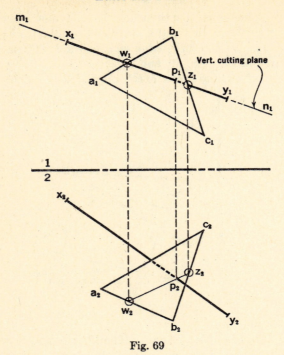

Fig. 69

4. Locate p_1 in the horizontal projection plane by projection.

5. Determine the visibility of line XY in both views by the method described in Fig. 64.

B. Edge View Method. Fig. 70 shows how the edge view method is used to find the piercing point in orthographic projection.

Example: Edge view method. (Refer to Fig. 70.)

Given: Horizontal and vertical projections of plane ABC and intersecting line XY.

Find: The point (P) of intersection of the line and the plane.

Procedure:

1. Find the edge view of the plane ABC (horizontal line is used in example).

2. View #3 shows plane ABC as an edge and line XY intersecting plane ABC at point P_3.

3. Project point P_3 back to the other views.

4. Determine visibility of line XY.

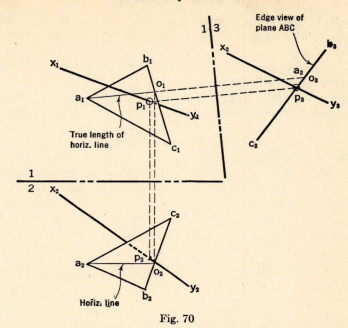

Fig. 70

37. Intersection of Two Planes. The intersection of two plane surfaces is a *straight line*. This line is determined by *two* points which are *common* to both intersecting planes. See **Fig. 71.**

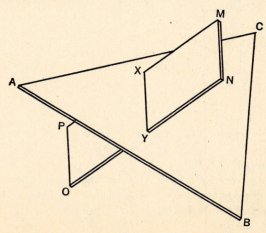

Fig. 71

Points X and Y are common to planes ABC and $MNOP$. The straight line connecting these points is the line of intersection of the two planes.

There are three main methods by which the line of intersection of two planes may be determined:

(a) Cutting plane or piercing point method, using vertical cutting planes or planes perpendicular to the vertical projection plane.

(b) Edge view method.

(c) Auxiliary cutting plane method, used for unlimited planes.

A. Cutting Plane or Piercing Point Method. The points of intersection (piercing points) of two lines of one of the intersecting planes are determined on the other plane by passing vertical cutting planes through these two lines. These two piercing points are then connected with a straight line, which is the required line of intersection. (Refer to Fig. 72.)

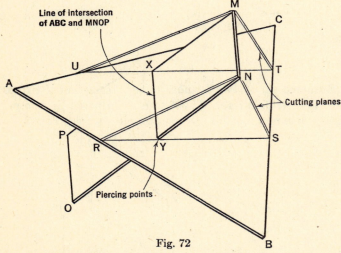

Fig. 72

Fig. 72 pictorially illustrates two cutting planes NRS and MTU which contain two lines (NO and MP) of the plane $MNOP$ and intersect the plane ABC.

The points where RS and TU, the lines of intersection of the cutting planes and the plane ABC, cross the lines NO and MP are the **piercing points** Y and X of the lines NO and MP with the plane ABC.

The line of intersection is the straight line connecting piercing points X and Y.

Fig. 73 shows an orthographic representation of the intersection of two planes.

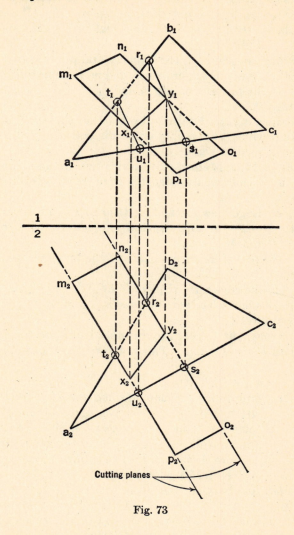

Fig. 73

Example: Cutting plane method.   (Refer to Fig. 73.)

Given: Horizontal and vertical projections of intersecting planes *ABC* and *MNOP*.

Find: The line of intersection *XY* of the two given planes.

Procedure:

1. Pass cutting planes perpendicular to the vertical projection plane containing lines n_2o_2 and m_2p_2 of plane $MNOP$ and intersecting plane ABC. These lines intersect plane $a_2b_2c_2$ at r_2s_2 and t_2u_2.

2. Project r_2s_2 and t_2u_2 to the horizontal projection plane.

3. Where r_1s_1 and t_1u_1 cross n_1o_1 and m_1p_1 respectively are the locations of the piercing points x_1 and y_1. Connect these points with a straight line.

4. Project x_1y_1 to the vertical projection plane to find x_2y_2.

5. Determine the visibility of the respective planes, using the method described in Fig. 64.

B. Edge View Method. In a view where one of the intersecting planes appears as an edge, the *edge view plane* is considered to be a cutting plane.

Example: Edge view method. (Refer to Fig. 74.)

Given: Horizontal and vertical projections of intersecting planes ABC and $MNOP$.

Find: The line of intersection XY of the two given planes.

Procedure:

1. Find the edge view of plane ABC.

2. Edge view $a_3b_3c_3$ crosses lines n_3o_3 and m_3p_3 at points x_3 and y_3 respectively.

3. Project x_3 and y_3 back to the other views and connect these points with a straight line. This is the required line of intersection.

4. Determine visibility.

C. Auxiliary Cutting Plane Method for Unlimited Planes. If auxiliary cutting planes intersect two given unlimited planes, the auxiliary planes cut lines on **each** of the given planes. The **points** at which these lines intersect are **common** to the given planes, and therefore they lie on the **line of intersection** of the two given planes. (See Fig. 75, p. 90.)

Example: Auxiliary cutting plane method. (Refer to Fig. 76, p. 91.)

Given: Horizontal and vertical projections of two unlimited and intersecting planes ABC and $MNOP$.

Find: The line of intersection of planes ABC and $MNOP$.

Procedure:

1. Draw auxiliary cutting planes I and II. (Vertical cutting

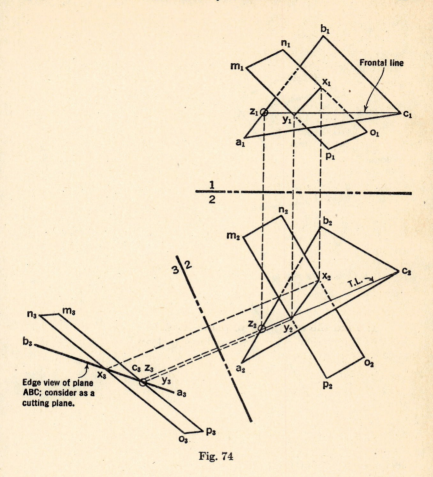

Fig. 74

planes are used in example. Any planes perpendicular to the horizontal and vertical projection planes can be used.)

2. Plane I cuts lines q_1r_1 and s_1t_1 on planes $a_1b_1c_1$ and $m_1n_1o_1p_1$. Plane II cuts lines u_1v_1 and w_1h_1 on planes $a_1b_1c_1$ and $m_1n_1o_1p_1$.

3. Project these lines to the vertical projection plane and extend them so that they intersect at x_2 and y_2.

4. Connect x_2 and y_2 with a straight line. This is the required line of intersection.

5. Project x_2 back to view #1, being careful that the pro-

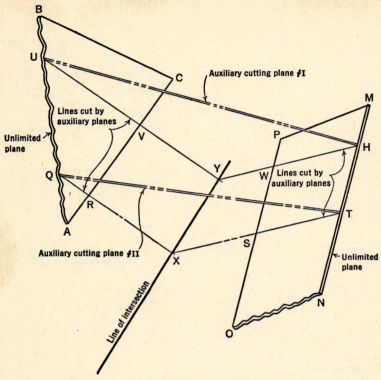

Fig. 75

jector stops at cutting plane #I (point x_1).

    6. Project y_2 back to view #1 onto cutting plane #II (point y_1).

    Note: The line of intersection in this case can be considered to be indefinite in extent.

    38. Dihedral Angle between Two Planes. Two intersecting planes form an angle which is called the *dihedral angle* (see Fig. 77).

    The *true dihedral angle* is seen in a view where the line of intersection of the two planes appears as a *point,* and therefore in this same view the planes appear as *edges.*

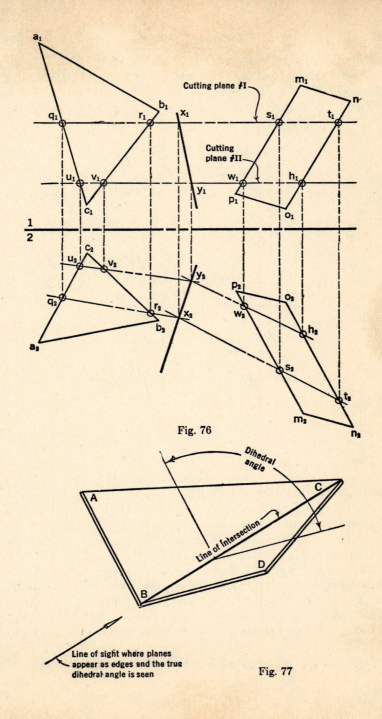

Fig. 76

Fig. 77

Dihedral angle

Line of Intersection

Line of sight where planes appear as edges and the true dihedral angle is seen

Example: Dihedral angle. (Refer to Fig. 78.)

Given: Horizontal and vertical projections of intersecting planes *ABC* and *BCD*.

Find: The true dihedral angle formed by planes *ABC* and *BCD*.

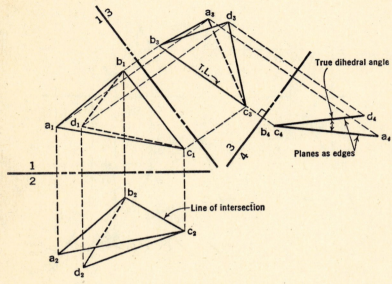

Fig. 78

Procedure:

1. Find the true length of the line of intersection. (Draw reference line 1–3 parallel to b_1c_1.)

2. Find line of intersection as a point. (Draw reference line 3–4 perpendicular to true length b_3c_3.)

3. In view #4 planes *ABC* and *BCD* appear as edges. The true dihedral angle is measured in this view.

39. Angle between a Line and a Plane. To see the *true angle* between any line and a plane, the plane must be seen as an *edge* and the line must appear in its *true length*. (See Fig. 79.)

Example: Angle between a line and a plane. (Refer to Fig. 80.)

Given: Horizontal and vertical projections of line *XY* and plane *ABC*.

Find: The true angle that line *XY* makes with plane *ABC*.

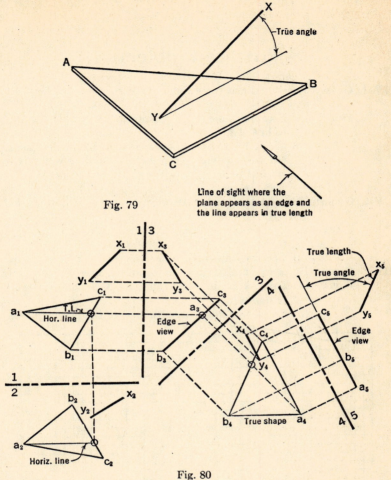

Fig. 79

Line of sight where the plane appears as an edge and the line appears in true length

Fig. 80

Procedure:

1. Find plane ABC as an edge (true length horizontal line seen as a point is used in example).

2. Find true shape of plane ABC (draw reference line 3–4 parallel to $a_3b_3c_3$).

3. Find true length of line XY (draw reference line 4–5 parallel to x_4y_4). View #5 shows both the edge view of plane ABC and the true length of line XY. The true angle is measured in this view.

Practice Problems

IV–1. Given: Horizontal and vertical projections of two skew lines AB and CD.

  Find: **(a)** Create a plane which contains line CD and is parallel to line AB.

   (b) Check by finding an edge view of the parallel plane. (In the edge view either both given lines will appear parallel or one line will appear as a point, proving that the constructed plane is parallel.)

IV–2. Given: Horizontal and vertical projections of two skew lines EF and GH.

  Find: **(a)** Draw a plane which contains line GH and is parallel to line EF.

   (b) Check by finding an edge view of the parallel plane.

IV–3. Given: Horizontal and vertical projections of lines AB and CD.

  Find: **(a)** Draw a plane which contains CD and is parallel to line AB.

   (b) Check by finding an edge view of the parallel plane.

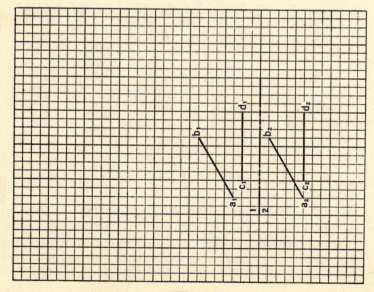

IV–1

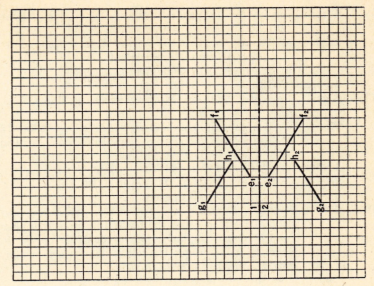

IV–2

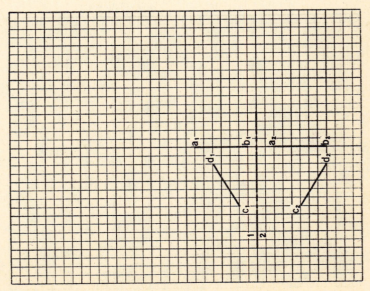

IV–3

IV-4. Given: Horizontal and vertical projections of line AB and point X.

Find: (a) The true length of the shortest distance from point X to line AB by the line method.

(b) Check result by using the plane method.

Answer: 1 inch.

IV-5. Given: Horizontal and vertical projections of line CD and point Y.

Find: (a) The true length of the shortest line from Y to line CD.

(b) The bearing and slope of the shortest line. Measure and note all data on drawing.

Answer: (a) T.L. of shortest line = $1\frac{1}{16}$ inches.

(b) Bearing from Y = due south.

Slope = 48 degrees.

IV-6. Given: Point E is located $\frac{1}{2}$ inch behind the vertical projection plane and $\frac{1}{2}$ inch below the horizontal projection plane. Point F is located $1\frac{1}{4}$ inches behind the vertical projection plane, $1\frac{1}{2}$ inches below the horizontal projection plane, and $1\frac{3}{4}$ inches to the right of E.

Find: Draw the horizontal and vertical projections of a point Z which is located $1\frac{3}{4}$ inches below the horizontal projection plane and 1 inch from the midpoint of a line connecting E and F. Show all construction in all views.

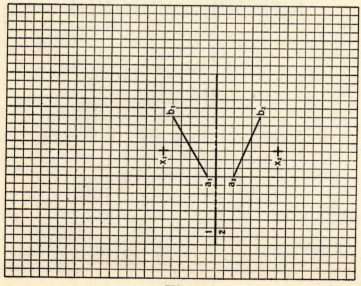

IV-4

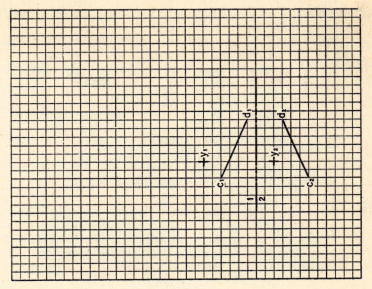

IV–5

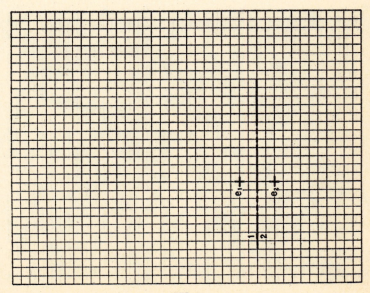

IV–6

IV-7. Given: Horizontal and vertical projections of two skew lines AB and CD.

Find: (a) The shortest distance (XY) between lines AB and CD by the line method.

(b) Draw the horizontal and vertical projections of XY.

Answer: T.L. of shortest distance $XY = 1\frac{1}{16}$ inches.

IV-8. Given: Horizontal and vertical projections of two skew lines EF and GH.

Find: (a) The shortest distance (LM) between EF and GH by the plane method.

(b) Draw the horizontal and vertical projections of LM.

Answer: T.L. of shortest distance $LM = \frac{5}{16}$ inch.

IV-9. Given: Horizontal and vertical projections of two skew lines AB and CD.

Find: (a) The shortest horizontal line XY between AB and CD.

(b) Draw the horizontal and vertical projections of XY.

Answer: T.L. of the shortest horizontal line $XY = \frac{13}{16}$ inch.

IV-7

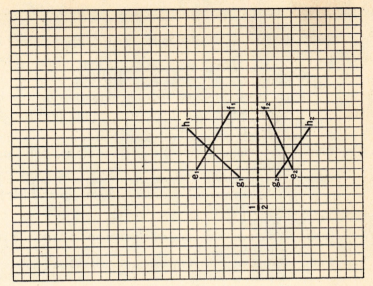

IV–8

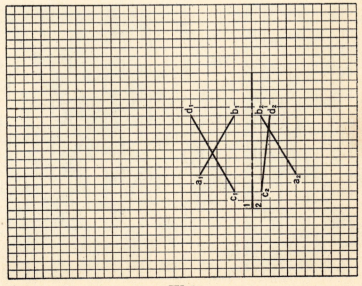

IV–9

IV–10. Given: Horizontal and vertical projections of two skew lines CD and EF.

  Find: **(a)** The shortest horizontal line OP between lines CD and EF.

   (b) The shortest line LM between CD and EF having a grade of 40% down from EF.

   (c) Draw the horizontal and vertical projections of OP and LM.

  Answer: **(a)** T.L. of $OP = \frac{7}{8}$ inch.

   (b) T.L. of $LM = \frac{3}{4}$ inch.

IV–11. Given: Horizontal and vertical projections of plane ABC and point P.

  Find: The true length of the shortest distance from point P to the plane ABC. Measure and note this true length on the drawing.

  Answer: $1\frac{5}{16}$ inches.

IV–12. Given: Horizontal and vertical projections of plane DEF and point Q.

  Find: The horizontal and vertical projections of the projection of point Q on the plane DEF. (The foot of a perpendicular line from Q to plane DEF is the projection of point Q on plane DEF.)

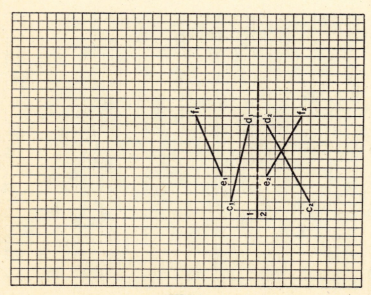

IV–10

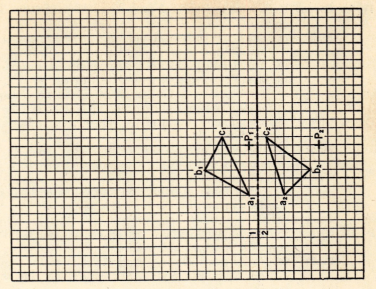

IV-11

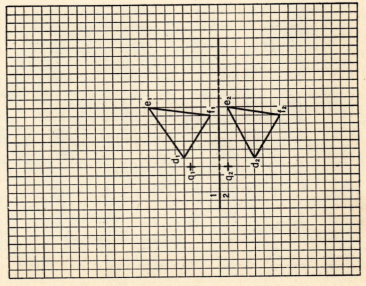

IV-12

IV–13. Given: Horizontal and vertical projections of plane *GHI* and point *R*.

   Find: **(a)** Draw the horizontal and vertical projections of a line *RS* from *R* perpendicular to plane *GHI*.

   (b) Through *S* (the foot of the perpendicular) draw any line *XY* in the plane *GHI*. Prove that *XY* is perpendicular to *RS* by drawing the true shape of the plane formed by *RS* and *XY*.

IV–14. Given: Horizontal and vertical projections of plane *ABC* and point *P*.

   Find: **(a)** The shortest distance from *P* to plane *ABC*. Measure and note the true length of this distance on the drawing.

   (b) Draw the horizontal and vertical projections of the shortest distance from *P* to plane *ABC*.

   Answer: Shortest distance = $1\frac{1}{8}$ inches.

IV–15. Given: Horizontal and vertical projections of two parallel lines *WX* and *YZ* and a point *Q*.

   Find: **(a)** The shortest distance from point *Q* to the plane formed by *WX* and *YZ*. Measure and note this distance on the drawing.

   (b) Draw the horizontal and vertical projections of the shortest distance from *Q* to plane *WXYZ*.

   Answer: Shortest distance = $1\frac{1}{4}$ inches.

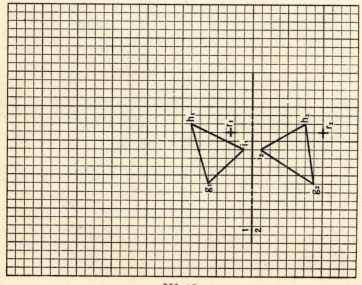

IV–13

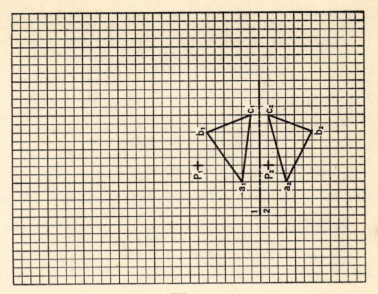

IV–14

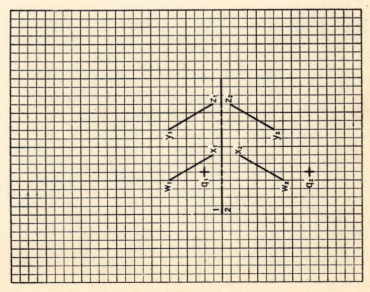

IV–15

IV–16. Given: Horizontal and vertical projections of plane *ABC* and line *MN*.

Find: (a) The piercing point of line *MN* and plane *ABC* by the edge view method.

(b) Show visibility of *MN* in all views.

IV–17. Given: Horizontal and vertical projections of plane *EFG* and line *XY*.

Find: (a) The piercing point of line *XY* and plane *EFG* by the vertical cutting plane method.

(b) Show visibility of line *XY* in all views.

(c) Check by edge view method.

IV–18. Given: Horizontal and vertical projections of lines *AB* and *MN*, and point *X*.

Find: (a) The piercing point of line *MN* and the plane formed by line *AB* and point *X*.

(b) Show visibility of *MN* in all views.

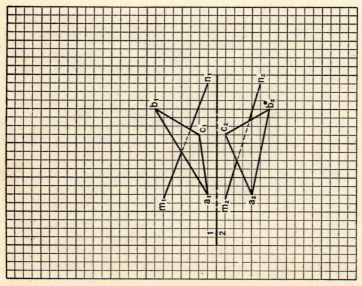

IV–16

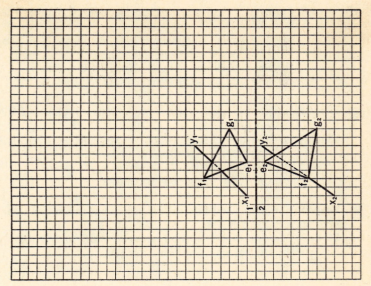

IV–17

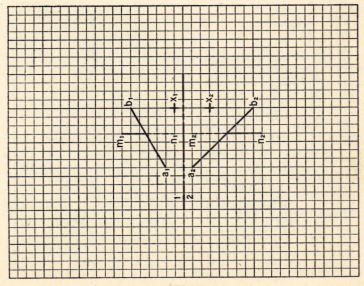

IV–18

IV–19. Given: Horizontal and vertical projections of planes *ABC*
and *WXY*.

Find: (a) The line of intersection between planes *ABC* and
WXY by the cutting plane method.

(b) Show visibility of both planes in all views.

(c) Solve by the edge view method for a check.

IV–20. Given: Horizontal and vertical projections of planes *LMN*
and *RSTU*.

Find: (a) The line of intersection between planes *LMN* and
RSTU by the piercing point method.

(b) Show visibility of both planes in all views.

(c) Check by the edge view method.

IV–21. Given: Horizontal and vertical projections of unlimited planes
ABC and *WXY*.

Find: The line of intersection between *ABC* and *WXY* by
the auxiliary cutting plane method.

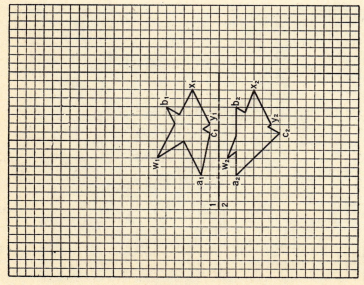

IV–19

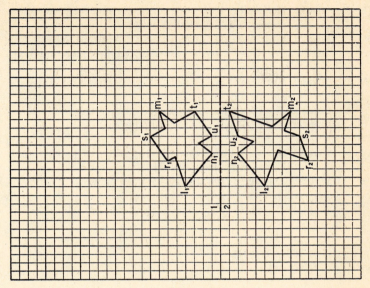

IV–20

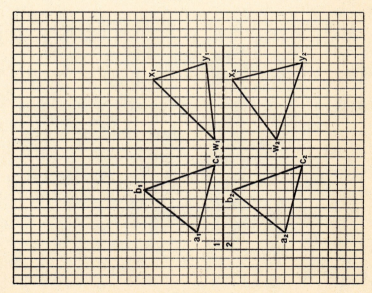

IV–21

IV–22. Given: Horizontal and vertical projections of two unlimited planes *LMNO* and *WXYZ*.

     Find: Find the line of intersection between planes *LMNO* and *WXYZ* by the auxiliary cutting plane method.

IV–23. Given: Horizontal and vertical projections of intersecting planes *ABC* and *ABD*.

     Find: The true dihedral angle between planes *ABC* and *ABD*.

     Answer: True dihedral angle = 21 degrees.

IV–24. Given: Horizontal and vertical projections of two intersecting planes *ABC* and *WXY*.

     Find: The true dihedral angle between planes *ABC* and *WXY*.

     Suggestion: First find line of intersection between the two given planes by the piercing point method.

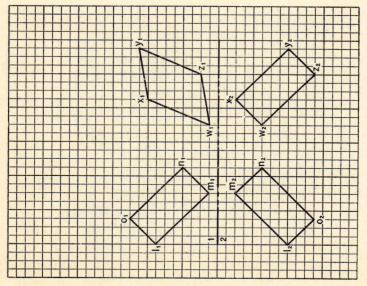

IV–22

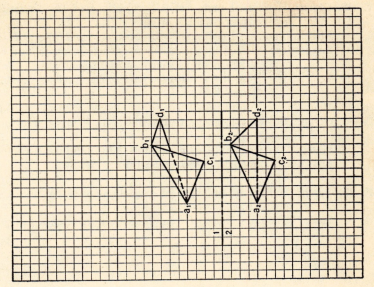

IV–23

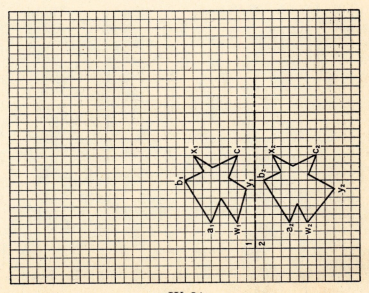

IV–24

IV-25. **Given:** Horizontal and vertical projections of two unlimited intersecting planes *ABC* and *PQR*.

Find: The true dihedral angle between planes *ABC* and *PQR*.

Suggestions: Use two sheets of cross-section paper and set up problem as shown on right-hand sheet. Find the line of intersection by the auxiliary cutting plane method.

IV-26. **Given:** Horizontal and vertical projections of plane *ABC* and line *XY*.

Find: The true angle between plane *ABC* and line *XY*.

Answer: True angle = 65 degrees (or 115 degrees).

IV-27. **Given:** Horizontal and vertical projections of plane *DEF* and line *XY*.

Find: The true angle between plane *DEF* and line *XY*.

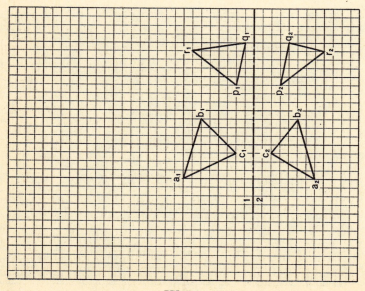

IV-25

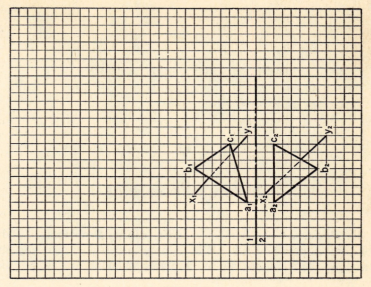

IV–26

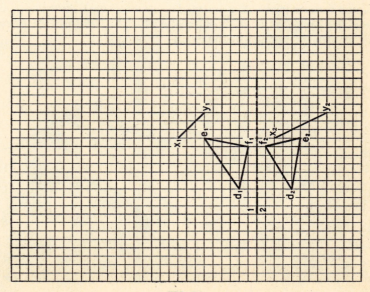

IV–27

V

Mining Problems

40. Principles Involved in Solving Mining Problems. In the solution of problems related to mining operations, basically the principles applied are those involving skew lines and parallel planes.

In order to undertake the solution of different types of mining problems, it is necessary to become familiar with some basic mining terminology. A *stratum of ore* (vein of ore), for descriptive geometry purposes, consists of a plane which has *uniform thickness* and is usually inclined in the earth's crust.

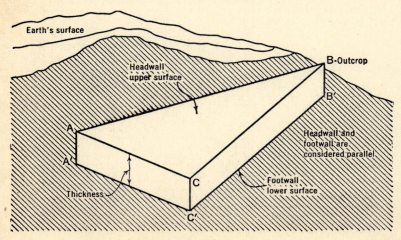

Fig. 81

41. Mining Terminology and Definitions. The following definitions relate to a stratum of ore.

In Fig. 81 ABC and $A'B'C'$ represent a *portion* of an ore vein.

Headwall. The top surface (ABC) of a stratum of ore.

Footwall. The lower surface ($A'B'C'$) of a stratum of ore.

Thickness. The perpendicular distance between the headwall and the footwall.

Outcrop. The point at which a section of the stratum of ore lies at or above the surface of the ground.

To determine the exact position of a stratum of ore, usually it is necessary to drill *bore holes* from the surface of the earth to the ore stratum. These bore holes may be vertical or inclined.

In addition, the *outcrop* can also be used as one point for determining the position of an ore stratum. (See Fig. 82.)

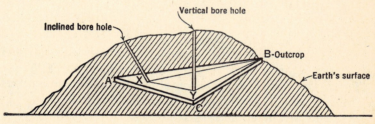

Fig. 82

Bore holes. Holes drilled from the earth's surface to a stratum of ore, used to determine the position of the stratum underground.

When bore holes are drilled, they intersect the ore stratum at various points. If *three* points are located on the surface of the stratum, these are sufficient to determine the plane of the ore stratum, since three points determine a plane. See Fig. 82; note plane XYB on the surface of the ore stratum.

When the *plane* of the stratum is determined, its position may be fixed by noting its *dip* and *strike*.

Dip. The angle an ore stratum makes with the horizontal projection plane (in other words, the *slope* of a plane).

Strike. The *bearing* of a level or *horizontal* line on the plane of the ore stratum (see Fig. 83, p. 114).

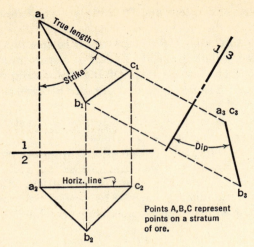

Fig. 83

42. Parallel Bore Hole Problem.

Example: Mining problem involving parallel bore holes and the outcrop of a vein of ore. (Refer to Fig. 84 and Fig. 85.)

Given: The vertical and horizontal projections of *three points* on a vein of ore. These points are determined as follows:

a. Point *B* is determined by the outcrop.

b. *Vertical* bore hole #I intersects the headwall at point *A* 50' below the earth's surface and the footwall at 150'.

c. *Vertical* bore hole #II intersects the headwall at point *C* 300' below the earth's surface and the footwall at 400'.

Bore hole #I is located 500' west and 250' south of point *B*.

Bore hole #II is located 600' south and 100' west of point *B*.

Find: The strike, dip, and thickness of the vein of ore.

Procedure: Refer to Fig. 85.

1. Strike. Draw a horizontal line in the plane $a_2b_2c_2$ and find the true length of this line in the horizontal projection plane. Strike is noted here as N 75° E.

2. Dip. See T.L. of the horizontal line as a point so that edge view $a_3b_3c_3$ of the vein of ore can be seen. Measure dip in view #3.

3. Thickness. Thickness is measured in view #3, where edge view of the vein of ore is seen.

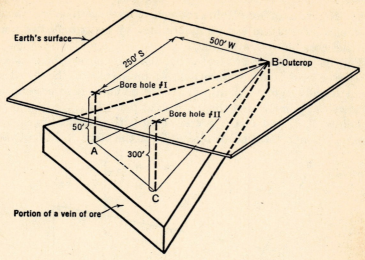

Fig. 84

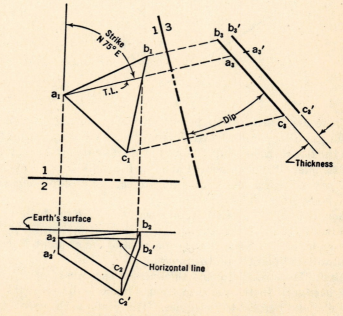

Fig. 85

43. Nonparallel Bore Hole Problem. The dip, strike, and thickness of a vein of ore can be determined by using *only two* nonparallel bore holes (see Fig. 86).

Fig. 86 pictorially illustrates two nonparallel bore holes intersecting a vein of ore. Bore hole #I intersects the headwall at point B and the footwall at point C. Bore hole #II intersects the headwall at point E and the footwall at point F.

Points E, B and F, C are connected with straight lines. EB lies in the plane of the headwall and FC in the plane of the footwall.

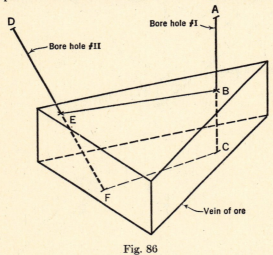

Fig. 86

There are two basic methods by which a mining problem involving two nonparallel bore holes may be solved, the *plane method* and the *line method*. Fig. 87 and Fig. 88 illustrate these methods.

Example: Plane method. (Refer to Fig. 87.)

Given: Vertical and horizontal projections of two nonparallel bore holes AC and DF.

Bore hole AC intersects a vein of ore at point B on the headwall and point C on the footwall.

Bore hole DF intersects the same vein of ore at point E on the headwall and point F on the footwall.

Find: The strike, dip, and thickness of the vein of ore.

Procedure: In the plane method a plane is drawn containing one line and parallel to the other. This plane is then seen as an edge.

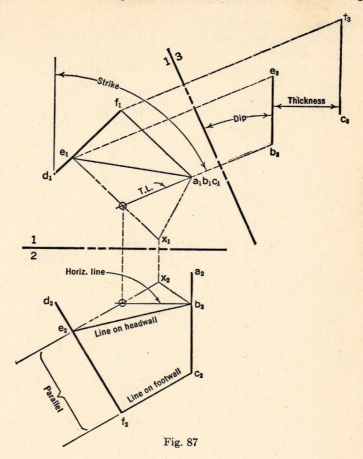

Fig. 87

1. Create a plane containing line *EB* and parallel to line *FC*. (**Assume** point x_2 and draw e_2x_2 parallel to f_2c_2.)

2. Draw e_1x_1 parallel to f_1c_1 (*EBX* is the required parallel plane).

3. Find plane *EBX* as an edge (horizontal line seen as a point is used in example). View #1 shows T.L. of horizontal line; therefore the **strike** can be noted here.

4. View #3 shows plane *EBX* as an edge. In this view lines e_3b_3 and c_3f_3 are parallel. The **dip** and **thickness** of the vein of ore are seen in this view.

Example: Line method. (Refer to Fig. 88.)

Given: Vertical and horizontal projections of two nonparallel bore holes *AC* and *DF*.

Find: The strike, dip, and thickness of the vein of ore.

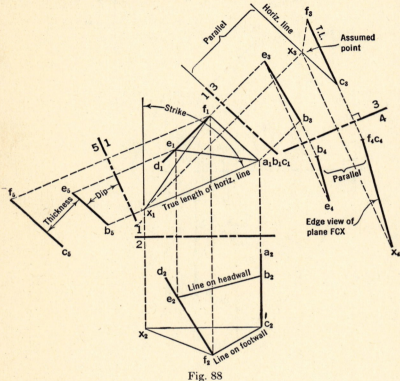

Fig. 88

Procedure:

1. Connect *EB* and *FC* with straight lines in both given views.

2. Find the *true length* of *FC* (draw reference line 1–3 parallel to f_1c_1).

3. From c_3 draw a *horizontal line* (parallel to reference line 1–3).

4. *Assume* point x_3 on this horizontal line and form plane $f_3c_3x_3$.

5. Find plane $f_3c_3x_3$ as an edge by drawing reference line 3–4 perpendicular to true length f_3c_3.

Note: Before x_4 can be located so that the position of the edge view of plane $f_3c_3x_3$ can be fixed, it is necessary to project line e_3b_3

to view #4. Since lines *EB* and *FC* lie in planes that are parallel to each other, plane *FCX* will be *parallel* to line e_4b_4 in view #4.

6. Locate x_1 and x_2 by projection. Measure **strike** in horizontal projection plane.

7. To find the dip of the vein of ore draw reference line 1–5 perpendicular to the true length of *XC* (x_1c_1). Note **dip** and **thickness** in view #5.

Practice Problems

V–1. Given: Horizontal and vertical projections of points *A*, *B*, and *C* on the headwall and points *D*, *E*, and *F* on the footwall of a vein of ore.

Find: (a) The dip of the vein.
(b) The strike of the vein.
(c) The thickness of the vein.

Suggestion: Use a scale of $\frac{1}{4}'' = 1'0''$ to measure thickness.

Answers: (a) Dip = 45 degrees.
(b) Strike = S 77° W.
(c) Thickness = 3′8″.

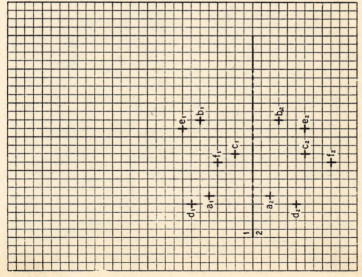

V–1

V-2. Given: Three points on a vein of ore are determined as follows:

    (a) Point *A* on the headwall is determined by the outcrop (horizontal and vertical projections given).

    (b) One vertical bore hole (#1) intersects the headwall at point *B* 400 feet below the surface of the earth and the footwall at 600 feet.

    (c) A second bore hole (#2) intersects the headwall at point *C* 200 feet below the surface of the earth and the footwall at 400 feet.

    (d) Bore hole #1 is located 600 feet north and 400 feet east of the outcrop *A*.

    (e) Bore hole #2 is located 100 feet south and 1000 feet east of the outcrop *A*.

Find: **(a)** The dip of the vein.

    (b) The strike of the vein.

    (c) The thickness of the vein.

Suggestion: Use a scale of $\frac{1}{4}'' = 100'$.

Answers: **(a)** Dip = 28.5 degrees.

    (b) Strike = N 63° E.

    (c) Thickness = 175'.

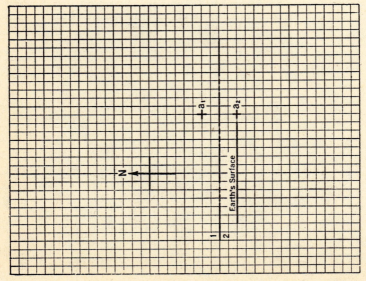

V-3. Given: Horizontal and vertical projections of points WXY on the headwall of a vein of ore and point P the starting point of a tunnel.

Find:
    (a) The shortest tunnel from P to the vein of ore.

    (b) Draw the horizontal and vertical projections of the shortest tunnel and measure and note the bearing and slope of this tunnel.

    (c) The shortest tunnel having a grade of 20% down from P to the vein of ore.

    (d) Draw the horizontal and vertical projections of this grade tunnel and measure and note the bearing of this tunnel.

Suggestions: Consider tunnel to be a straight line. Use a scale of $\frac{1}{4}'' = 100'$.

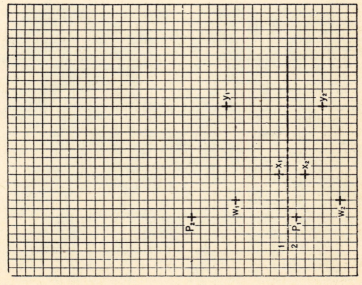

V-3

V-4. Given: Two nonparallel bore holes are drilled in the following manner: Bore hole #1 starting at point *A* is drilled so that it has a bearing of S 30° E and a slope of 45°. It intersects the headwall of a vein of ore after drilling 150 feet and the footwall after drilling 600 feet.

Bore hole #2 starting at point *D* has a bearing of S 20° E and a slope of 50°. It strikes the headwall at 250 feet and the footwall at 700 feet.

Find: Use the plane method to determine the following data:
- (a) The strike of the vein of ore.
- (b) The dip of the vein of ore.
- (c) The thickness of the vein of ore.

Suggestion: Use a scale of ¼″ = 100′.

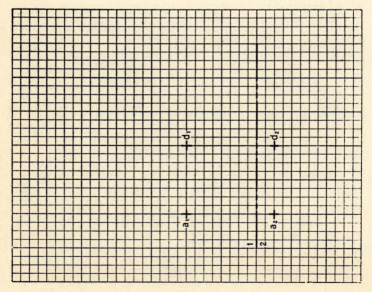

V–4

V-5. Given: Same data as problem #V-4.

  Find: Using the line method determine the strike, dip, and thickness of the vein of ore. Compare results with the plane method solution.

Suggestion: Use a scale of $\frac{1}{4}'' = 100'$.

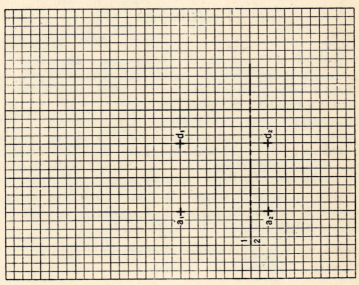

V-5

VI

Revolution

44. Position of Observer.   Up to this point in the solution of descriptive geometry problems the *observer* "moved" to various positions in order to see the true length of a line, the true shape of a plane, the edge view of a plane, etc.

These same problems can be solved more quickly by applying the method of revolution.   In this method the observer *remains* in a *stationary* position while the line, plane, or object is *revolved* so that its true length or shape, etc. presents itself to the observer.

45. Revolution of a Point about an Axis.   When a point is rotated about an axis it moves at a fixed distance from the axis. The *path* of the rotated point is a *circle* which has its plane *perpendicular* to the *axis* (see Fig. 89 and Fig. 90).

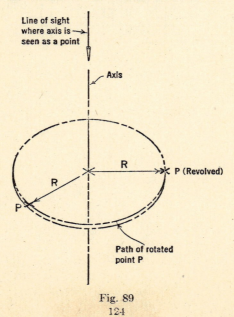

Fig. 89

124

Fig. 90 shows the rotation of point P about an axis in orthographic projection.

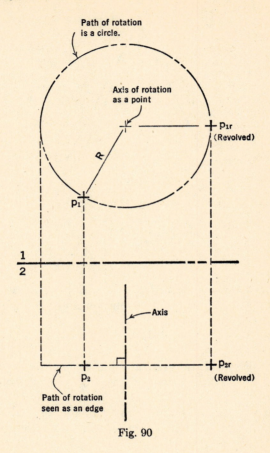

Fig. 90

46. Revolution of a Line about an Axis.   When a line is revolved about another line as an *axis* a cone of revolution is generated (see Fig. 91).

The **true length** and **slope** of a line can be found by applying the principles of revolution illustrated in Fig. 91.   In order for line AB to be seen in true length it must be revolved so that it will be perpendicular to the line of sight of the observer.   (Note the two positions of line AB where the true length can be seen.)

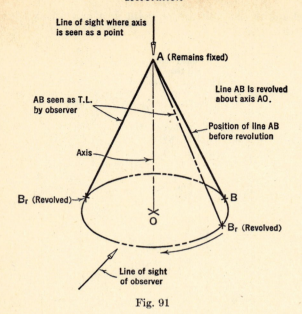

Line of sight where axis
is seen as a point

A (Remains fixed)

Line AB is revolved
about axis AO.

AB seen as T.L.
by observer

Position of line AB
before revolution

Axis

B_r (Revolved)

B

O

B_r (Revolved)

Line of sight
of observer

Fig. 91

Fig. 92 shows how the true length and slope of a line are found by applying the principles of revolution and orthographic projection.

Example: True length and slope of a line by revolution. (Refer to Fig. 92.)

Given: Horizontal and vertical projections of line AB.

Find: True length and slope of line AB by revolution.

Procedure:

Revolve a_1b_1 so that it is parallel to the vertical projection plane. True length and slope are seen in the vertical projection plane.

In Fig. 92 line AB was revolved about point A *on* the line. If necessary, line AB can be revolved about *any* point either on the line or outside the line.

Example: Revolution of a line about an axis outside the line. (Refer to Fig. 93, p. 128.)

Given: Horizontal and vertical projections of line AB and axis XY.

Find: True length and slope of line AB by revolving AB about XY.

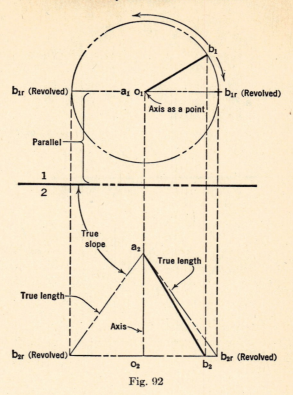

Fig. 92

Procedure:

1. In top view where axis XY appears as a point draw a line from the axis perpendicular to line a_1b_1 (at point p_1).

2. From axis x_1y_1 swing arcs to indicate the direction of rotation of points p_1, a_1 and b_1 (to the right in the figure).

3. Draw a line ***parallel*** to the vertical projection plane and tangent to the arc on which point p_1 was rotated.

4. The tangent point is p_{1r} (revolved).

5. Where the parallel line intersects the arcs on which a_1 and b_1 were rotated are the locations of a_{1r} and b_{1r}.

6. Project the revolved points to the vertical projection plane, where the true length and slope of line AB are seen.

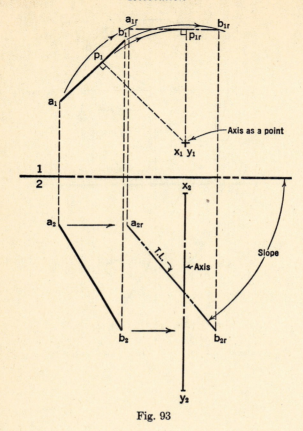

Fig. 93

Note: Points a_2 and b_2 moved directly to the right and parallel to the horizontal projection plane.

47. Revolution of a Plane about an Axis. A plane can be revolved about a line as an axis. The axis can lie in the plane or outside the given plane.

Fig. 94 illustrates pictorially how the true shape of a plane can be seen by revolving the plane so that it lies *parallel* to a projection plane (or *perpendicular* to the line of sight of an observer).

Example: True shape and slope of a plane by revolution. (Refer to Fig. 95.)

Given: Horizontal and vertical projections of plane *ABC*.

Find: True shape and slope of plane *ABC* by revolution.

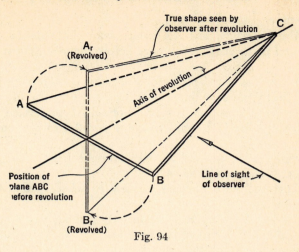

True shape seen by observer after revolution

A_r (Revolved)

C

A

Axis of revolution

Position of plane ABC before revolution

B

Line of sight of observer

B_r (Revolved)

Fig. 94

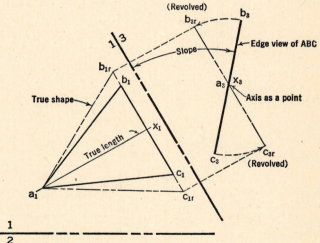

(Revolved)

b_{3r} b_3

1   3

Slope   ← Edge view of ABC

b_{1r} b_1

a_3 x_3

Axis as a point

True shape

True length x_1

c_3 c_{3r} (Revolved)

a_1 c_1

c_{1r}

1
―
2

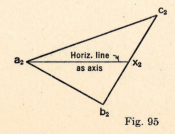

c_2

a_2 Horiz. line → x_2 as axis

b_2 Fig. 95

Procedure:

1. Use horizontal line AX as an axis. AX must be seen as a point and plane ABC as an edge before plane ABC can be revolved.

2. View #3 shows the edge view of plane ABC revolved about a_3x_3 *parallel* to reference line 1–3.

3. Project b_{3r} and c_{3r} to the horizontal projection plane. Note that in view #1 b_{1r} and c_{1r} move *parallel* to reference line 1–3.

4. Connect a_1, b_{1r}, and c_{1r} with straight lines. This is the required true shape of plane ABC.

5. The true slope appears in view #3 where plane ABC is seen as an edge.

48. Dihedral Angle by Revolution. By passing a cutting plane *perpendicular* to the line of intersection of two given planes and

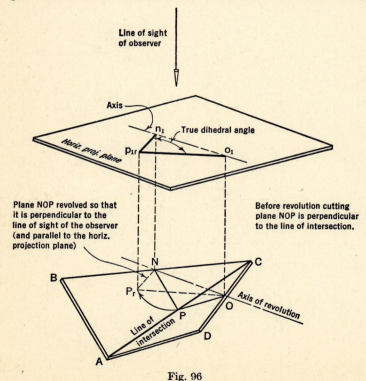

Fig. 96

revolving the cutting plane so that it is ***perpendicular*** to the line of sight of the observer, the true dihedral angle will be seen. (See Fig. 96.)

Example: True dihedral angle by revolution. (Refer to Fig. 97.)

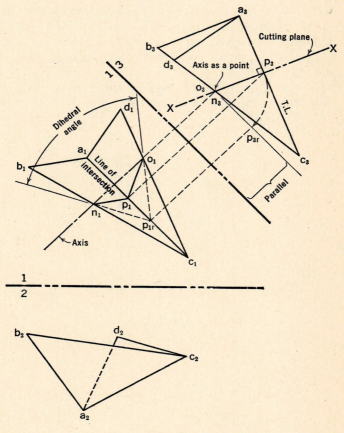

Fig. 97

Given: Horizontal and vertical projections of intersecting planes *ABC* and *ACD*.

Find: The true dihedral angle formed by the planes *ABC* and *ACD* using the method of revolution.

Procedure:

1. Find the true length of the line of intersection (draw reference line 1–3 parallel to a_1c_1).

2. Draw cutting plane X–X perpendicular to T.L. line a_3c_3. Plane X–X cuts given planes at p_3, o_3, and n_3. Project these points to view #1.

3. Revolve section $p_3o_3n_3$ about o_3n_3 as an axis parallel to the horizontal projection plane (parallel to reference line 1–3).

4. Locate p_1 revolved (p_{1r}) in view #1 by projection.

5. Connect p_{1r} to n_1 and o_1 with straight lines. The true dihedral angle is $n_1p_{1r}o_1$.

Note: Section $p_3o_3n_3$ can be revolved about any point in view #3. Solve the example by revolving about p_3.

Practice Problems

VI-1. Given: Horizontal and vertical projections of line AB and point P.

Find: (a) Revolve point P backward 90° about line AB as an axis.

(b) Show revolved position of P in both given views.

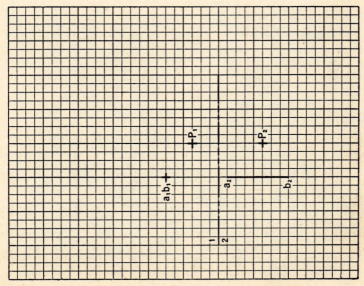

VI-1

VI–2. **Given:** Horizontal and vertical projections of line XY and point Q.

     Find: (a) Revolve point Q counterclockwise 270 degrees about XY as an axis.

          (b) Show revolved position of point Q in all views.

VI–3. **Given:** Horizontal and vertical projections of line AB.

     Find: (a) The true length of AB by revolving AB about b_1.

          (b) Check T.L. by revolving AB about b_2.

VI–4. **Given:** Horizontal and vertical projections of lines LM and CD.

     Find: The true length of line CD by revolving it about LM as an axis.

     Answer: T.L. of $CD = 2\frac{1}{8}$ inches.

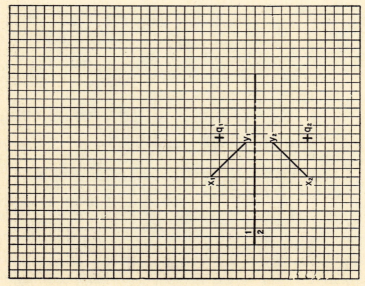

VI–2

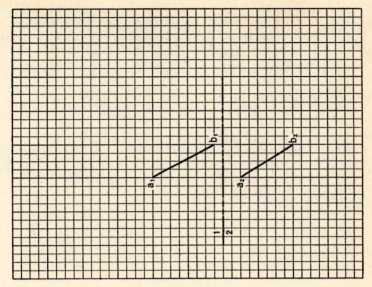

VI–3

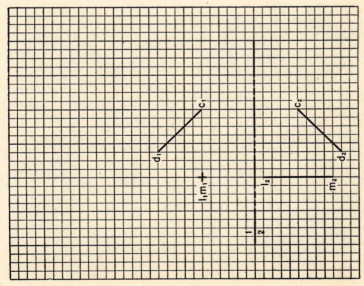

VI–4

VI-5. Given: Horizontal and vertical projections of lines AB and CD.
   Find: (a) Revolve line CD about AB as an axis so that its true length appears in an elevation view.
   (b) Show revolved position of CD in all views.
Answer: T.L. of $CD = 1\frac{11}{16}$ inches.

VI-6. Given: Horizontal and vertical projections of plane ABC.
   Find: The true shape of plane ABC by revolution.

VI-7. Given: Horizontal and vertical projections of plane DEF.
   Find: (a) Revolve plane DEF so that its true shape appears in view #2.
   (b) Measure and note the true angle FDE.
Answer: Angle $FDE = 75$ degrees.

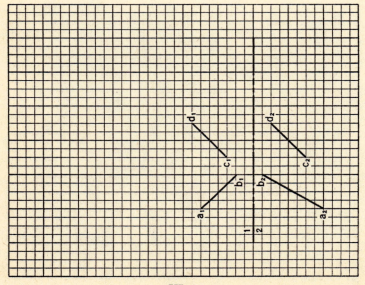

VI-5

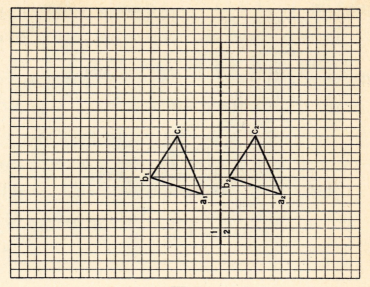

VI-6

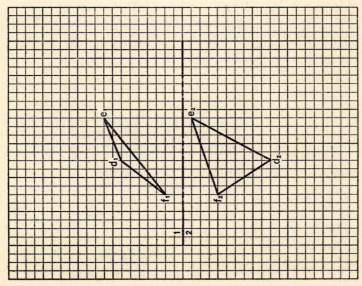

VI-7

VI–8. Given: Horizontal and vertical projections of intersecting planes *ABC* and *ABD*.

   Find: The dihedral angle between the given planes by revolution.

   Answer: Dihedral angle = 15 degrees.

VI–9. Given: Horizontal and vertical projections of intersecting planes *EFG* and *EFH*.

   Find: (a) The dihedral angle between the given planes by revolution.

   (b) Check by drawing a view where line of intersection is seen as a point.

VI–10. Given: Horizontal and vertical projections of two unlimited intersecting planes *LMN* and *WXY*.

   Find: (a) The dihedral angle between the given planes by revolution.

   (b) Check by drawing a view where the line of intersection appears as a point.

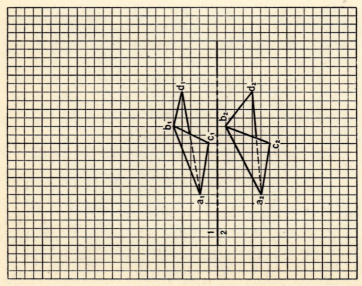

VI–8

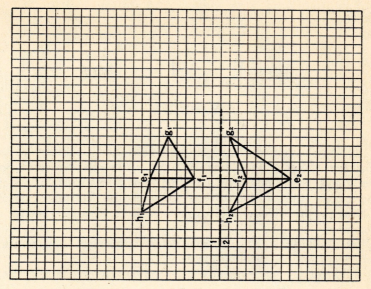

VI–9

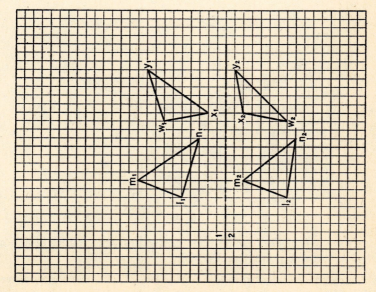

VI–10

VII

Vectors[*]

49. Application of Principles of Descriptive Geometry to the Solution of Vector Problems. A force can be represented graphically as a vector quantity. If concurrent noncoplanar forces act on a body, the solution of these forces can be determined by applying the principles of orthographic projection used in descriptive geometry.

50. Definitions. The following definitions pertain to the solution of concurrent noncoplanar forces.

Vector. A line segment which represents a force, acceleration, or velocity. This line has magnitude, sense, direction, and position (see Fig. 98).

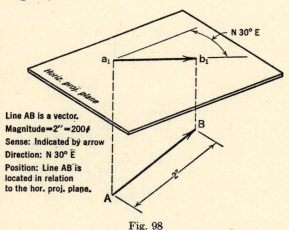

Fig. 98

Concurrent Forces. Forces (or vectors) whose lines of action intersect at one point (see Fig. 99).

* Vector scales indicated in this chapter are not full size owing to picture reductions.

Coplanar Forces. Forces (or vectors) whose lines of action all lie in the same plane (see Fig. 100).

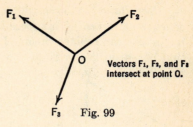

Vectors F₁, F₂, and F₃ intersect at point O.

Fig. 99

Noncoplanar Forces. Forces (or vectors) whose lines of action do *not* lie in the same plane (see Fig. 101).

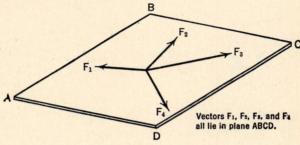

Vectors F₁, F₂, F₃, and F₄ all lie in plane ABCD.

Fig. 100

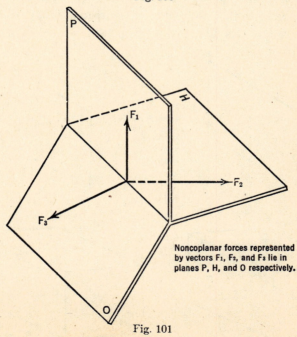

Noncoplanar forces represented by vectors F₁, F₂, and F₃ lie in planes P, H, and O respectively.

Fig. 101

Resultant. A force which is a combination of two or more concurrent forces. This force is the equivalent of the concurrent forces and can replace them. It has magnitude, sense, and direction based on the original concurrent forces. (See Fig. 102 and Fig. 103.)

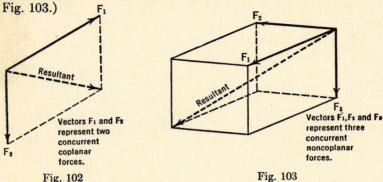

Vectors F_1 and F_2 represent two concurrent coplanar forces.

Fig. 102

Vectors F_1, F_2 and F_3 represent three concurrent noncoplanar forces.

Fig. 103

Equilibrant. The force which is necessary to balance or keep in equilibrium two or more concurrent forces. The equilibrant is *equal* in value to the resultant but its direction is *opposite* to that of the resultant (see Fig. 104 and Fig. 105).

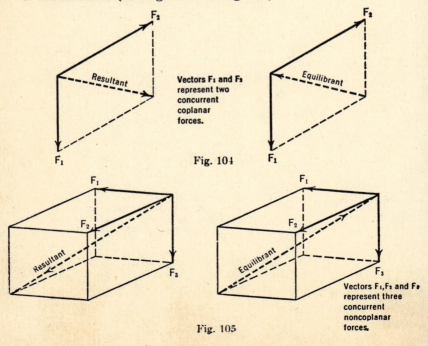

Vectors F_1 and F_2 represent two concurrent coplanar forces.

Fig. 104

Vectors F_1, F_2 and F_3 represent three concurrent noncoplanar forces.

Fig. 105

Structure Diagram. An actual scale drawing of a loaded structure (see Fig. 106).

Space Diagram. A diagram *not* drawn to scale which shows the *direction* and *sense* of the forces in a loaded structure (see Fig. 107).

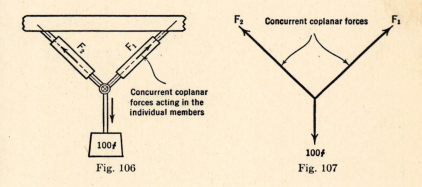

Fig. 106   Fig. 107

Vector Diagram. A diagram in which vectors are drawn to *scale* in *consecutive* order and *parallel* to the forces represented in the space diagram (see Fig. 108).

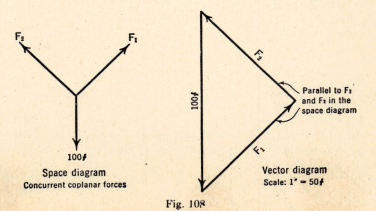

Fig. 108

In order to draw space and vector diagrams of concurrent *noncoplanar* forces it is necessary to show at least *two* orthographic views of the forces since *noncoplanar* forces lie in more than one plane. See Fig. 109 and Fig. 110. (Fig. 109 pictorially illustrates three concurrent, noncoplanar forces F_1, F_2, and F_3.)

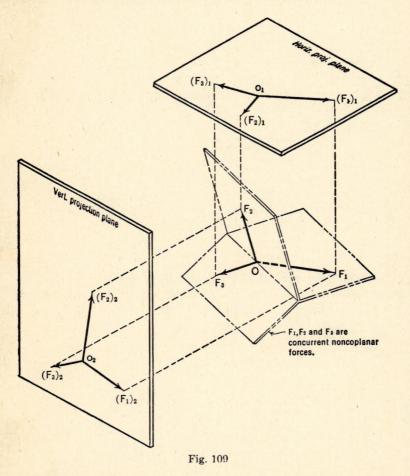

Fig. 109

Fig. 110 shows the horizontal and vertical projections (in orthographic projection) of the space and vector diagrams of forces F_1, F_2, and F_3.

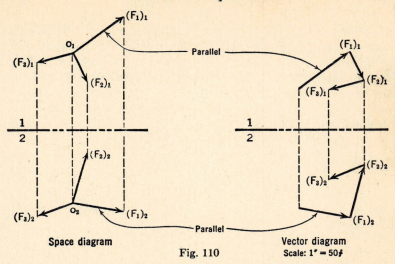

Space diagram   Vector diagram

Fig. 110   Scale: 1″ = 50#

51. Application of Descriptive Geometry Principles to Find the Resultant and Equilibrant of Concurrent Noncoplanar Forces.

Example: Resultant and equilibrant. (Refer to Fig. 111.)

Given: Space diagram showing the horizontal and vertical projections of three concurrent noncoplanar forces: $F_1 = 120\#$; $F_2 = 115\#$; and $F_3 = 80\#$.

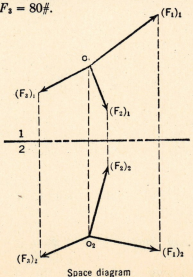

Fig. 111

Space diagram

Find: The resultant and equilibrant of the given forces.

In order to draw the vector diagram the lengths of the horizontal and vertical *projections* of the vectors in the vector diagram must be determined. To do this it is necessary to find *true length lines* in the vector diagram upon which the *actual scale values* of the forces can be laid off. Fig. 112 and Fig. 112a illustrate two methods by which this can be accomplished (force F_1 is used in examples).

A. Conventional Method: Refer to Fig. 112.

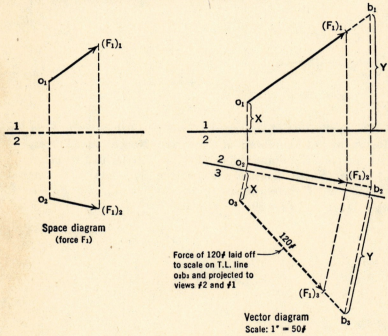

Space diagram
(force F_1)

Force of 120# laid off
to scale on T.L. line
o_3b_3 and projected to
views #2 and #1

Vector diagram
Scale: 1″ = 50#

Fig. 112

Procedure:

1. Starting from any point o_1 in the vector diagram draw line o_1b_1 (any convenient length) parallel to $o_1 (F_1)_1$, and from o_2, any convenient distance from reference line 1–2, draw o_2b_2 parallel to $o_2(F_1)_2$.

2. Find the true length of *OB* by passing reference line 2–3 parallel to o_2b_2.

3. Line o_3b_3 is true length, and on this line the actual scale value of force F_1 can be laid off ($F_1 = 120\#$).

4. Project scale length of F_1 back to views #2 and #1. This determines the length of the horizontal and vertical projections of the vector representing F_1. This process is repeated for all the given forces.

B. Rapid Method *: Refer to Fig. 112a.

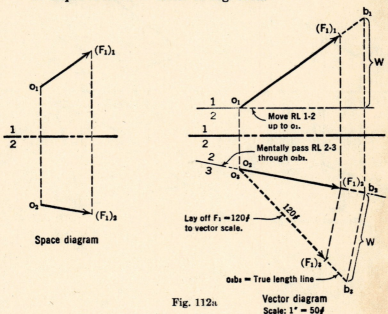

Space diagram

Fig. 112a

Vector diagram
Scale: 1″ = 50#

Procedure:

1. Starting from any point o_1 in the vector diagram draw line o_1b_1 (any convenient length) parallel to $o_1(F_1)_1$, and draw o_2b_2 parallel to $o_2(F_1)_2$.

2. Find the true length of OB by moving reference line 1–2 up to o_1 and mentally passing reference line 2–3 through o_2b_2 (note distance W).

3. Line o_3b_3 is true length, and on this line the actual scale value of force F_1 can be laid off.

4. Project scale length of F_1 back to views #2 and #1. This determines the length of the horizontal and vertical projections of the vector representing F_1. Repeat this process for all the given forces.

* Based on a presentation which appears on pages 73 and 74, and Fig. 65, in *Graphic Methods for Solving Problems* by Frank A. Heacock, Professor of Graphics, Princeton University, Copyright 1952 by F. A. Heacock.

Completed Solution of Forces Presented in Fig. 111: Refer to Fig. 112b.

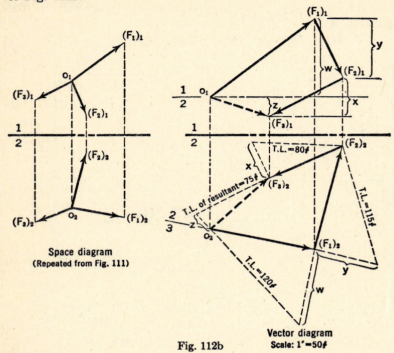

Space diagram
(Repeated from Fig. 111)

Fig. 112b

Vector diagram
Scale: 1' = 50#

Procedure:

1. Draw the vector diagram to scale (Rapid method used in example — all construction lines not shown to avoid confusion).

2. Connect o_2 (starting point of the vector chain) and $(F_3)_2$ (end of vector chain) with a straight line. This line closes the vector diagram and represents the vertical projection of the *resultant* since its direction is indicated as being opposite to that of the given forces.

3. Connect o_1 and $(F_3)_1$ with a straight line. This is the horizontal projection of the resultant.

4. Find the true length of the resultant by the rapid method. The value of the resultant is determined by the given scale.

5. The bearing and slope of the resultant can be found.

6. The equilibrant will be equal in value to the resultant but its sense is *opposite* to that of the resultant.

52. Tripod Problems: Loaded Structures Involving Concurrent Noncoplanar Forces.

A. Vertical Load. Fig. 113 shows a tripod structure supporting a vertical load of 100#. The 100# load is distributed among the three legs of the tripod as indicated by the arrows. Since the legs of the tripod are *not* of equal length and make different angles with the horizontal plane, the distribution of the 100# load in the individual legs is *not* equal.

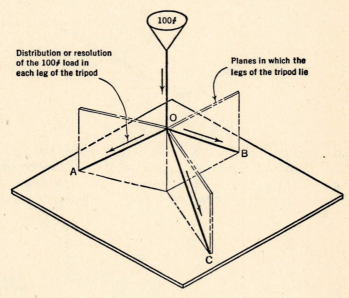

Fig. 113

In order for the tripod to be in equilibrium, the individual legs must jointly resist the external load of 100#. This resistance is in the form of *internal* forces (stresses) in the individual legs which *oppose* the external force. (See Fig. 114, p. 150.)

The legs of the tripod in Fig. 114 are in *compression;* that is, they must *push* to withstand the 100# load. Compressive stresses are denoted by arrows *acting toward* a common point (see Fig. 114 and Fig. 115, p. 150).

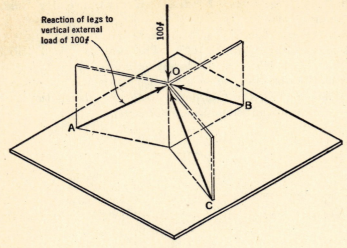

Fig. 114

If the 100# load in Fig. 114 were a vertical **upward** or pulling load, the individual legs of the tripod would be in **tension** since they would have to **pull against** the external load. Tensile stresses are denoted by arrows **acting away from** a common point (see Fig. 116).

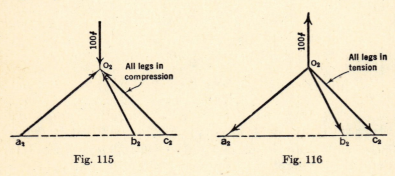

Fig. 115   **Fig. 116**

B. Horizontal Load. Fig. 117 shows a tripod which has a 100# **horizontal** load applied to it.

Note: Stress in leg *OC* is **pulling** against the 100# load; therefore it is in **tension** (since the arrow indicating the stress in *OC*

is going *away from* the common
center). Legs *OA* and *OB* are
pushing against the 100# load;
therefore these legs are in *compression* (since the arrows indicating the stresses in *OA* and *OB*
go *towards* the common center).

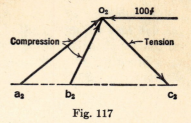

Fig. 117

C. Inclined Load. Apply analysis used in Fig. 117.

**53. General Method of Obtaining Stress Solutions in Tripods
Having Various Types of Loads Act-
ing upon Them.** When two views of
a space diagram involving a loaded
tripod are given, the general procedure used to solve for the stresses in
the legs of the tripod is as follows:

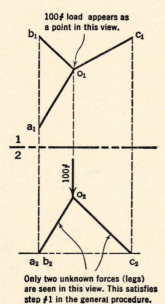

100# load appears as
a point in this view.

Only two unknown forces (legs)
are seen in this view. This satisfies
step #1 in the general procedure.

Fig. 118

(a) Look at the space diagram
so that the *known* load and *only two
unknown* forces (legs) are seen.

(b) Draw the *vector diagram.*

(c) Find and measure the true
lengths of the vectors.

(d) Transfer the direction of the
arrows (indicating the sense of the
stresses) to the space diagram. With
the arrows on the space diagram it is
possible to determine which legs are
in tension or compression.

Figs. 119 and 127 illustrate two
examples where two unknown forces
are seen one behind the other in order
to satisfy step (a) in the above pro-
cedure. In these examples the plane of two unknown forces is
seen as an edge.

Example: The plane of two unknown forces seen as an edge.
(Refer to Fig. 118 and Fig. 119.)

Given: Horizontal and vertical projections of a space diagram
showing a vertically loaded tripod where the plane of two legs of
the tripod is seen as an edge.

Find: The stresses in each leg of the tripod, indicating the type
of stress that is acting in each leg (tension or compression).

Determination of Vector Sequence in the Vector Diagram through Bow's Notation (Modified Form). Bow's notation is a method which facilitates the drawing of a vector diagram by setting up a procedure to fix the sequence of the vectors in the vector diagram.

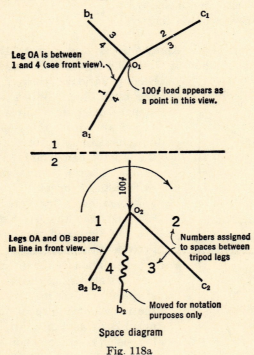

Space diagram

Fig. 118a

Bow's Notation Procedure: Refer to Fig. 118a.

1. In one view of the given space diagram, represent one of the legs of the tripod which appears in line with another leg by a zigzag line (o_2b_2 used in example).

2. In this same view assign a ***number*** to each ***space*** between the legs (including zigzag o_2b_2) and the 100# load. *Always* start numbering on either side of the ***known load,*** then proceed to the single leg (o_2c_2), and *lastly* number the two legs which appear in line with each other.

3. Draw a curved arrow to indicate the sequence in which the forces or vectors are to be laid off in the vector diagram. (Clockwise direction is used in example.)

4. In the adjacent view (top view in example) of the space diagram indicate the *numbers* that are on *either side* of the legs in the original view (front view).

5. The sequence in which the vectors will be drawn in the vector diagram is now fixed.

Fig. 118a shows the space diagram with Bow's notation applied. The next step is to draw the actual vector diagram by starting with the *known load* and then proceeding to lay off the vectors in the direction indicated by the curved arrow (see Figs. 119 to 125 inclusive). The vector representing the 100# load begins at #1 and ends at #2, next vector representing stress in *OC* begins at #2 and ends at #3, etc.

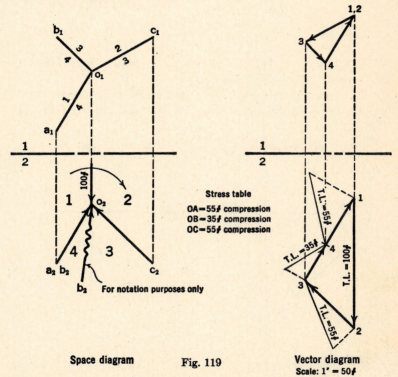

Stress table

OA = 55# compression
OB = 35# compression
OC = 55# compression

For notation purposes only

Space diagram   Fig. 119   Vector diagram
Scale: 1″ = 50#

Complete Stress Solution: Refer to Fig. 119, and to Figs. 120–125 for details.

Procedure for Vector Diagram Layout: Refer to Figs. 119 to 125 inclusive.

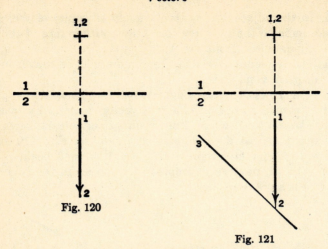

Fig. 120

Fig. 121

1. Draw vector 1–2 parallel to the 100# load in the front view. Show vector 1–2 as a point in the top view (Fig. 120).

2. From end of vector 1–2 draw the *direction* of vector 2–3 (parallel to o_2c_2). The end of vector 2–3 is not known at this stage (Fig. 121).

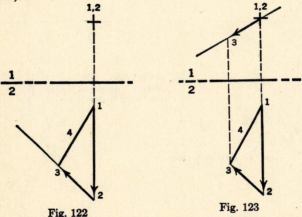

Fig. 122

Fig. 123

3. From starting point of vector 1–2 draw *direction* of vectors 3–4 and 4–1 (parallel to o_2a_2 and o_2b_2); these two forces have the same direction in the front view. The intersection of the line indicating the direction of vector 2–3 and the line indicating the direction of vectors 3–4 and 4–1 is the end of vector 2–3 (Fig. 122).

4. To determine the end of vector 3–4 it is necessary to complete the top view. From vector 1–2 (which appears as a point) draw vector 2–3 (parallel to o_1c_1). The end of this vector is determined by projection from the front view (Fig. 123).

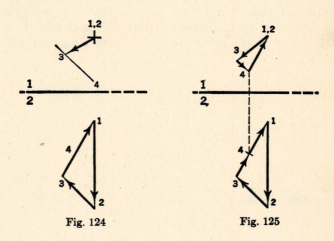

Fig. 124   Fig. 125

5. From the end of vector 2–3 draw direction of vector 3–4 (parallel to o_1b_1) (Fig. 124).

6. From the point view of vector 1–2 draw direction of vector 4–1 (parallel to o_1a_1). The point of intersection of the last two lines drawn is the location of the end of vector 3–4. Project this end (4) to front view (Fig. 125).

Note: The sense of all the arrows depends upon the sense of the known load. Since the loaded tripod is in equilibrium, the arrows *must follow each other* in the same direction and the vector diagram *must be a closed polygon.*

Upon completion of the vector diagram find the true lengths of the vectors by the rapid method, and transfer the direction of the arrows from the vector diagram to the space diagram. Note tension and compression forces as the case warrants. (See Fig. 119 for completed solution.)

54. Stresses in the Legs of a Tripod Determined by Viewing the Plane of Two Legs as an Edge. If two views of a loaded tripod are given so that no two legs of the tripod are seen in line with each other, it is necessary to obtain another view where this condition can be satisfied.

Fig. 126 shows pictorially the line of sight of the observer where the plane of two legs of the given tripod appears as an edge. (Two intersecting lines form a plane; therefore intersecting legs *OA* and *OB* form a plane. See Sec. 25 and Fig. 46, Chap. III, for edge view of a plane.)

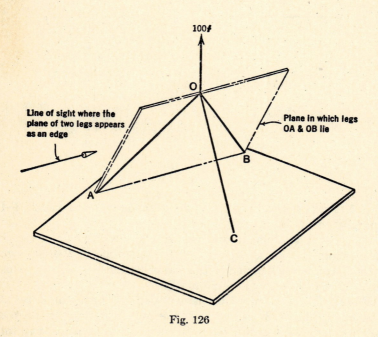

Fig. 126

Example: The plane of two unknown forces seen as an edge. (Refer to Fig. 127.)

Given: Horizontal and vertical projections of a space diagram showing a vertically loaded tripod.

Find: The stresses in each leg of the tripod. Note type of stresses.

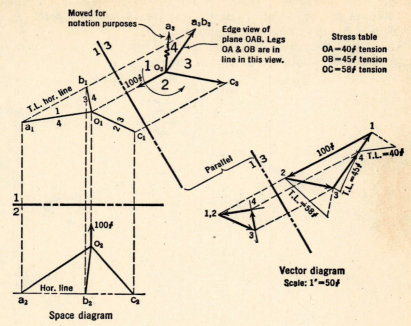

Fig. 127

Procedure:

1. Find plane *AOB* (which is formed by legs *OA* and *OB*) as an edge. (Horizontal line seen as a point is used in example.)

2. In view #3 where legs *OA* and *OB* appear in line apply Bow's notation, starting with the known force and continuing in the direction of the circular arrow.

3. Draw the vector diagram to scale. (Note proper sequence and directions of vectors.)

4. Find and measure the true lengths of the individual vectors using the rapid method.

5. Transfer arrows from the vector diagram to the space diagram. Since all the arrows act *away from* the common point *O*, the legs of the tripod are in tension.

55. Stress Solution Where One Unknown Force Is Seen as a Point. In order to satisfy the condition that known loads and *only two unknown forces* are to be seen in the space diagram before a stress solution (dealing with a loaded tripod) can be undertaken, **one** of the **unknown** forces can be seen as a *point* — thereby fulfilling the above requirement.

Fig. 128 illustrates pictorially a tripod with an **inclined load** applied to it. Note line of sight of observer.

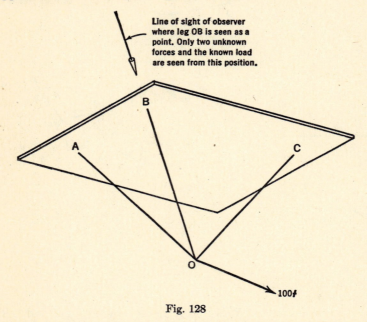

Fig. 128

Example: Stress solution in a loaded tripod where one unknown force is seen as a point. (Refer to Fig. 129.)

Given: Horizontal and vertical projections of a space diagram showing a tripod with an inclined load of 100# (= *OF*).

Find: The stresses in each leg of the tripod. Note type of stresses.

Procedure:

1. Find leg *OB* as a point. (Draw reference line 1–3 parallel to o_1b_1 to find the T.L. of *OB*. Then draw reference line 3–4 perpendicular to T.L. o_3b_3.)

2. In view #4 where *OB* appears as a *point* apply Bow's notation, starting with the known force and continuing in the direction of the circular arrow.

3. Draw the vector diagram to scale. Since the known load *OF* is an **inclined** force, its true length will **not** appear in the vector

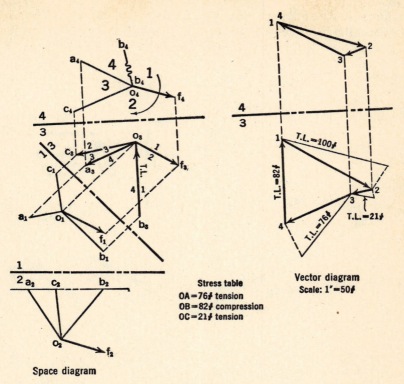

Stress table

OA = 76# tension
OB = 82# compression
OC = 21# tension

Vector diagram
Scale: 1″ = 50#

Space diagram

Fig. 129

diagram. By applying the rapid method explained in Fig. 112 for finding true lengths, the end points of *OF* can be fixed in the vector diagram.

4. Find and measure the true lengths of the individual vectors using the rapid method.

5. Transfer arrows from the vector diagram to the space diagram. Forces in legs *OA* and *OC* act away from the common point *O*; therefore these legs are in *tension*. The force in leg *OB* acts towards the common point *O*; therefore *OB* is in *compression*.

56. Resolution of Two Unknown Forces into One Unknown Force. By combining two unknown forces in a tripod into *one unknown* force a view can be obtained where only two unknown forces and the known load will be seen. With this condition established a stress solution can be undertaken. (See Fig. 130.)

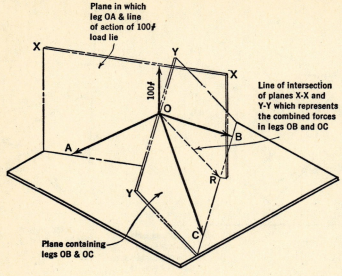

Plane in which leg OA & line of action of 100# load lie

100#

Line of intersection of planes X-X and Y-Y which represents the combined forces in legs OB and OC

Plane containing legs OB & OC

Fig. 130

Example: Stress solution involving the resolution of two unknown forces into one. (Refer to Fig. 131.)

Given: Horizontal and vertical projections of a space diagram showing a vertically loaded tripod.

Find: Stresses in each leg of the tripod. Note type of stresses.

Procedure:

1. Draw plane X–X through leg OA and the given 100# load (see view #1).

2. Find the line of intersection (OR) between plane X–X and the plane OBC in which the legs OB and OC lie. Line OR represents the combined forces acting in the legs OB and OC. The requirement of two unknown forces (OA and OR) and a known force (100#) is established.

3. Solve for OR (F_R) by drawing vector diagram #1. True lengths in this example were found by using the method of *revolution.*

4. Referring back to the space diagram, draw the edge view of the plane OBC (horizontal line seen as a point is used in example).

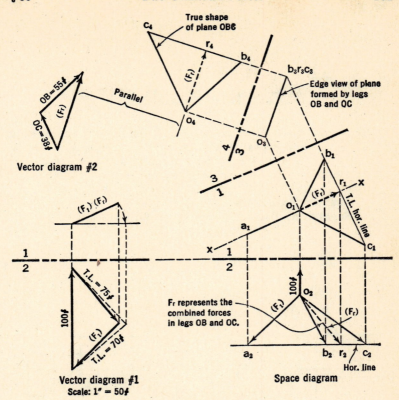

Fig. 131

5. Draw the true shape ($o_4b_4c_4$) of plane *OBC*.

6. Solve for forces in *OB* and *OC* by drawing vector diagram #2, using the value of F_R found in vector diagram #1.

7. Transfer arrows from vector diagrams to space diagram to determine the types of stresses involved.

Note: In this stress solution both vector diagrams dealt with *coplanar* forces, owing to the fact that two unknown forces were resolved into one unknown force.

Practice Problems

VII-1. Given: Space diagram showing the horizontal and vertical projections of four concurrent, noncoplanar forces: $F_1 = 100\#$, $F_2 = 80\#$, $F_3 = 120\#$, and $F_4 = 130\#$.

Find: (a) The resultant of the given forces.

(b) Show the resultant in the given space diagram.

Answer: Resultant = 115#.

Suggestion: Use two sheets of paper for this solution.

Vector scale: $\frac{1}{4}'' = 10\#$.

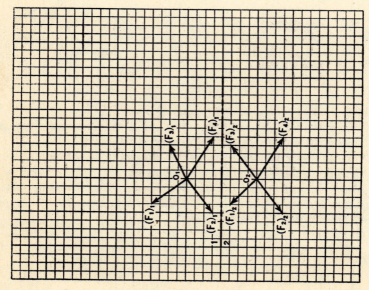

VII-1

VII-2. Given: Space diagram showing the horizontal and vertical
projections of a tripod (legs *OA*, *OB*, and *OC*) with a
100# vertical load applied to the tripod.

Find: The magnitude and type of stress that occurs in each
leg of the given tripod.

Answer: Stress in *OA* = 70# Tension.
Stress in *OB* = 30# Tension.
Stress in *OC* = 50# Tension.

Suggestion: Use two sheets of paper for this solution.
Vector scale: $\frac{1}{4}'' = 10\#$.

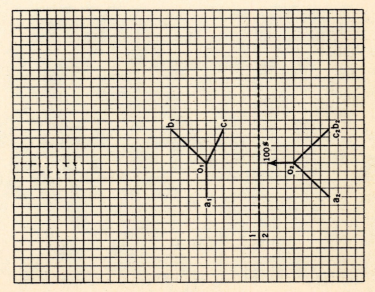

VII-2

VII-3. Given: Space diagram showing the horizontal and vertical projections of a tripod (legs OA, OB, and OC) with a 150# vertical load applied to the tripod.

Find: The magnitude and type of stress that occurs in each leg of the given tripod.

Answer: Stress in OA = 107.5# Compression.

Stress in OB = 85# Compression.

Stress in OC = 37.5# Compression.

Suggestion: Use two sheets of paper for this solution.

Vector scale: $\frac{1}{4}'' = 10\#$.

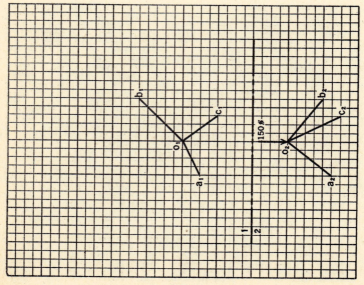

VII-3

VII-4. Given: Space diagram showing the horizontal and vertical projections of a tripod (legs *OX*, *OY*, and *OZ*) with an inclined load of 100# applied to the tripod.

Find: (a) The magnitude and type of stress which occurs in each leg of the given tripod by viewing the plane of two legs as an edge.

(b) Reproduce problem and solve by resolving two unknown forces into one. Compare answers for both solutions.

Suggestion: Use two sheets of paper for this solution.

Vector scale: $\frac{1}{2}'' = 10\#$.

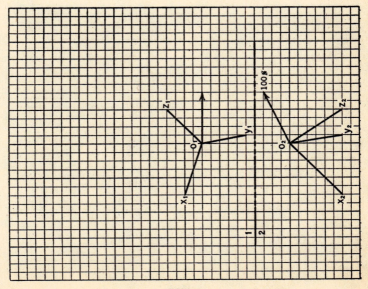

VII-4

VII-5. Given: Space diagram showing the horizontal and vertical projections of a tripod (legs OA, OB, and OC) with an inclined load of 120# applied to the tripod.

Find: The magnitude and type of stress which occurs in each leg of the given tripod by viewing one leg as a point.

Answer: Stress in OA = 145# Tension.

Stress in OB = 80# Compression.

Stress in OC = 8# Compression.

Suggestion: Use two sheets of paper for this solution.

Vector scale: $\frac{1}{4}''$ = 10#.

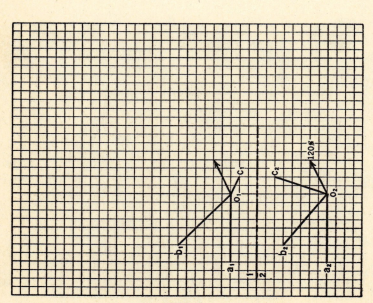

VII-5

VII-6. Given: Space diagram showing the horizontal and vertical projections of a tripod (legs OA, OB, and OC) with two inclined loads of 90# and 50# applied to the tripod.

Find: The magnitude and type of stress which occurs in each leg of the given tripod.

Answer: Stress in $OA = 16\#$ Compression.

Stress in $OB = 120\#$ Tension.

Stress in $OC = 10\#$ Tension.

Suggestions: Resolve the two inclined loads into one load. Use two sheets of paper for this solution.

Vector scale: $\frac{1}{4}'' = 10\#$.

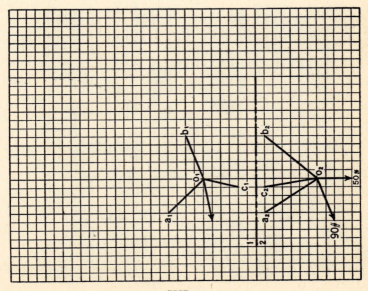

VII–6

VII-7. Given: Space diagram showing the horizontal and vertical
projections of a tripod (legs *OA*, *OB*, and *OC*) with
an 80# load applied to the tripod. The 80# load is
acting in a direction which is parallel to the profile
projection plane and makes an angle of 30° with the
vertical projection plane.

Find: (a) The magnitude and type of stress which occurs
in each leg of the given tripod by viewing the
plane of two legs as an edge.

  (b) Reproduce the problem and solve by viewing one
unknown force as a point. Compare answers for
the two solutions.

Suggestion: Use two sheets of paper for these solutions.
Vector scale: ¼″ = 10#.

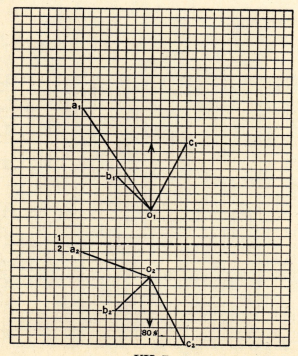

VII–7

VIII

Solids and Surfaces

57. Polyhedrons. Polyhedrons are multisided solids whose faces consist of intersecting plane surfaces. These solids are classified as regular or irregular polyhedrons. Figs. 132, 133, and 133a show a few examples of different types of polyhedrons.

Regular Polyhedrons: Faces of the solid *are* congruent.

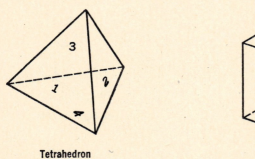

Tetrahedron
4 Faces

Hexahedron (Cube)
6 Faces

Fig. 132

Irregular Polyhedrons: Faces of the solid *are not* congruent.

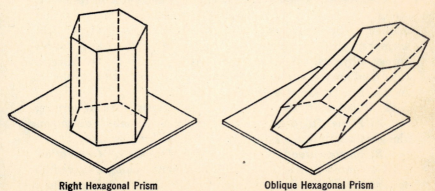

Right Hexagonal Prism

Oblique Hexagonal Prism

Fig. 133

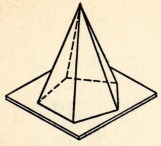

Right Pentagonal Pyramid

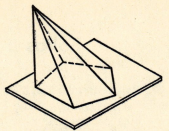

Oblique Pentagonal Pyramid

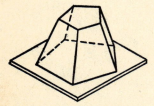

Frustum of a Right Pyramid

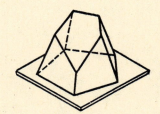

Truncated Right Pyramid

Fig. 133a

58. Surfaces. Geometrical surfaces are generated or formed by straight or curved lines which move through space in a prescribed manner. These surfaces are divided into two basic groups: *ruled surfaces,* which are generated by a *straight line,* and *double-curved surfaces,* which are generated by a *curved line.* Certain ruled and curved surfaces are classified as surfaces of revolution, such as the surface of a cone or a sphere.

59. Ruled Surfaces. There are three types of ruled surfaces:

A. Plane Surfaces. Generated by a straight line moving along 2 parallel lines or 2 intersecting lines (see Fig. 134).

B. Single-curved Surfaces (Two Methods of Generation). (1) Generated by a straight line which moves along a *curved* line so that each successive position (of the straight line) is parallel to its previous position (see Fig 135).

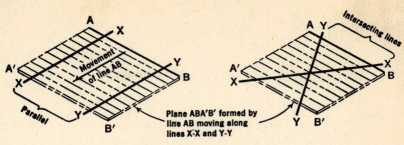

Plane ABA'B' formed by line AB moving along lines X-X and Y-Y

Fig. 134

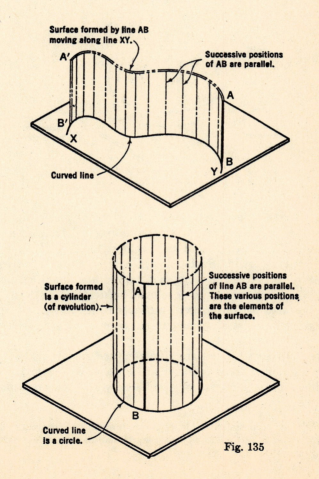

Surface formed by line AB moving along line XY.

Successive positions of AB are parallel.

Curved line

Surface formed is a cylinder (of revolution).

Successive positions of line AB are parallel. These various positions are the elements of the surface.

Curved line is a circle.

Fig. 135

(2) Generated by a straight line which is fixed at one end while the free end moves along a curved line (see Fig. 136).

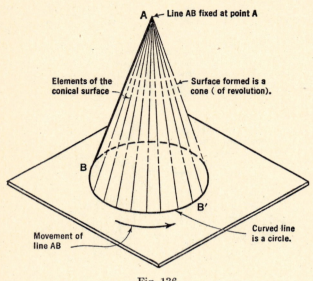

Fig. 136

C. Warped Surfaces.   Generated by a straight line which moves along 2 skew lines and remains parallel to a given plane (see Fig. 137).   They cannot be developed* accurately.

60. Double-curved Surfaces.   Double-curved surfaces are formed by revolving the *plane* of a curved line about an axis (see Fig. 138).   They cannot be developed accurately.

61. Surfaces of Revolution.   Surfaces of revolution are formed by *rotating* a line (straight or curved) about an axis (see Fig. 139 on p. 174).

Note: Surfaces formed are *single-curved* surfaces.

Many double-curved surfaces are surfaces of revolution.   Fig. 138 shows the generation of a sphere by revolving a semicircle about an axis.

* A surface which can be developed is one in which two consecutive lines in the surface will lie in the same plane when the surface is unrolled onto a flat plane.

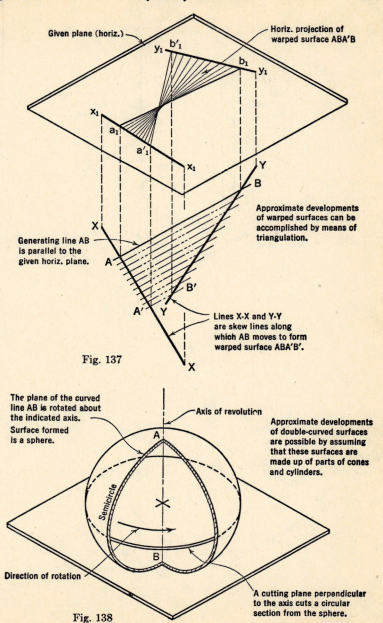

Given plane (horiz.)

Horiz. projection of
warped surface ABA'B

Approximate developments
of warped surfaces can be
accomplished by means of
triangulation.

Generating line AB
is parallel to the
given horiz. plane.

Lines X-X and Y-Y
are skew lines along
which AB moves to form
warped surface ABA'B'.

Fig. 137

The plane of the curved
line AB is rotated about
the indicated axis.
Surface formed
is a sphere.

Axis of revolution

Approximate developments
of double-curved surfaces
are possible by assuming
that these surfaces are
made up of parts of cones
and cylinders.

Semicircle

Direction of rotation

A cutting plane perpendicular
to the axis cuts a circular
section from the sphere.

Fig. 138

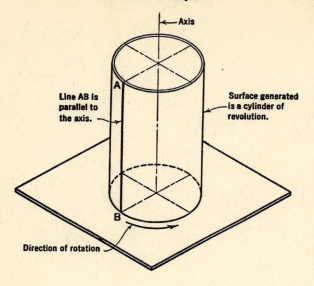

Axis

A

Line AB is parallel to the axis.

Surface generated is a cylinder of revolution.

B

Direction of rotation

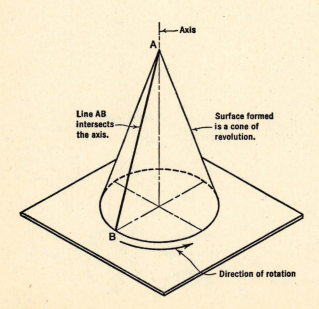

Axis

A

Line AB intersects the axis.

Surface formed is a cone of revolution.

B

Direction of rotation

Fig. 139

62. Points on a Prism. Before dealing with the topic of the intersections of surfaces, it is necessary to illustrate the principles involved in locating points on different surfaces, and also to show how the points of intersection of a line and a surface are determined.

Fig. 140a pictorially illustrates an assumed point P on an oblique rectangular prism. When the horizontal and vertical projections of the oblique prism are given and only the horizontal projection of point P is known, Fig. 140b shows how the vertical projection of point P is located. In Fig. 140a line XY through point P is determined by a vertical cutting plane containing P and intersecting the face $ABCD$. In Fig. 140b line x_1y_1 can be considered as the edge view of a vertical cutting plane or as any line through point P and in the plane $ABCD$ (see Sec. 24, pp. 52–53).

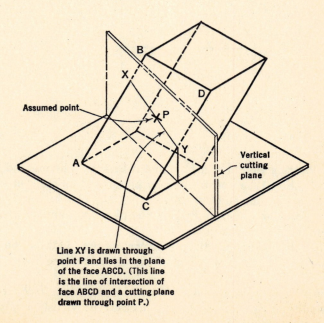

Assumed point

Vertical cutting plane

Line XY is drawn through point P and lies in the plane of the face ABCD. (This line is the line of intersection of face ABCD and a cutting plane drawn through point P.)

Fig. 140a

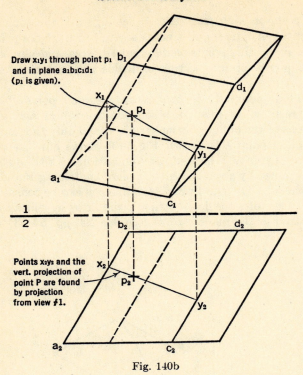

Draw x_1y_1 through point p_1
and in plane $a_1b_1c_1d_1$
(p_1 is given).

Points x_2y_2 and the
vert. projection of
point P are found
by projection
from view #1.

Fig. 140b

63. Piercing Points of a Line and a Prism. The method illustrated in Fig. 140a and b can also be applied to find the piercing points of a line and a prism.

Fig. 141a shows pictorially how the piercing points of line XY are found on the given prism.

The orthographic solution is shown in Fig. 141b. The horizontal and vertical projections of the prism and the line XY are given.

Points l_2, m_2, n_2, o_2, and p_2 are found by projection from view #1.

Visibility is determined by the method previously explained. Follow reasoning in detail.

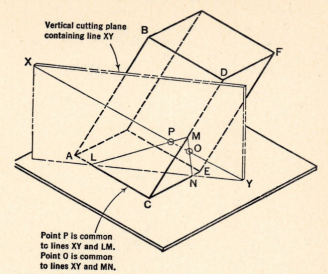

Vertical cutting plane
containing line XY

Point P is common
to lines XY and LM.
Point O is common
to lines XY and MN.

Fig. 141a

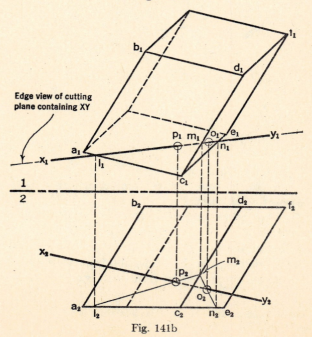

Edge view of cutting
plane containing XY

Fig. 141b

64. Points on a Cone.   In order to locate points on a cone it is necessary to utilize the *elements** of the cone. Fig. 142a and b illustrates the principle involved in fixing the location of a given point *P*.

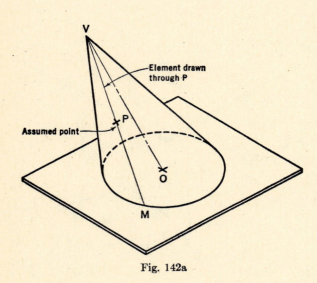

Fig. 142a

Example: Points on a cone.   (Refer to Fig. 142b.)

Given: Horizontal and vertical projections of cone *VO* and vertical projection of point *P*.

Find: The horizontal projection (p_1) of point *P*.

Procedure:

    1. Draw element v_2m_2 through p_2.

    2. Locate element v_1m_1 by projection from view #2.

    3. Locate p_1 by projection from view #2. Point p_1 lies on v_1m_1.

* An element is one position of a straight line which is used to generate a surface. (See Chap. VI, Fig. 91 and Fig. 92.)

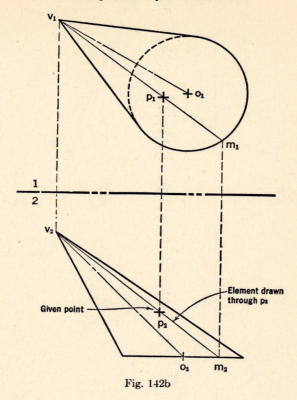

Fig. 142b

65. Piercing Points of a Line and a Cone. A cutting plane containing the ***vertex*** of a cone and ***intersecting the base*** cuts ***straight-line elements*** on the cone. If the cutting plane also contains a line which intersects the cone, the ***piercing points*** of the line will lie on the elements cut by the plane (see Fig. 143).

Example: Piercing points of a line and a cone. (Refer to Fig. 144.)

Given: Horizontal and vertical projections of cone *VO* and intersecting line *AB*.

Find: The piercing points of line *AB* on the surface of the cone.

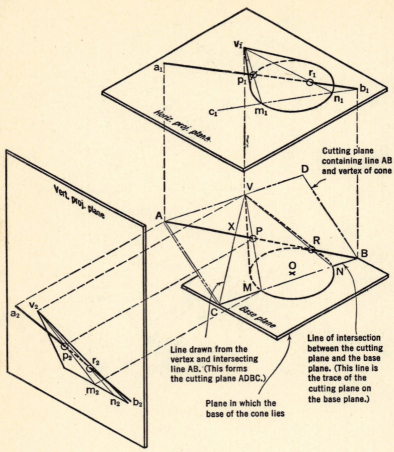

Cutting plane containing line AB and vertex of cone

Line of intersection between the cutting plane and the base plane. (This line is the trace of the cutting plane on the base plane.)

Line drawn from the vertex and intersecting line AB. (This forms the cutting plane ADBC.)

Plane in which the base of the cone lies

Fig. 143

Procedure: Refer to Fig. 144.

1. Extend line AB in view #2 so that it intersects the base plane at b_2.

2. Draw any line from v_2 which intersects a_2b_2 (at x_2) and the base plane at c_2. (Two intersecting lines form a plane.)

3. Find x_1, c_1, and b_1 by projection. Line c_1b_1 is the ***trace*** of the cutting plane on the base plane. It cuts the base of the cone at points m_1 and n_1.

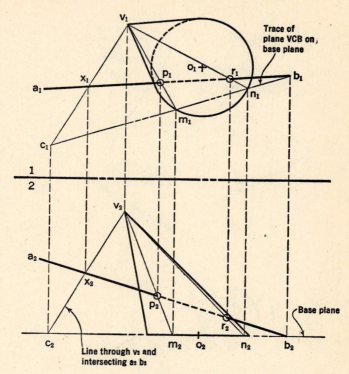

Fig. 144

4. Draw elements v_1m_1 and v_1n_1. The points at which these elements intersect a_1b_1 are the required piercing points (p_1 and r_1).

5. Locate p_2 and r_2 by projection.

6. Determine visibility of AB.

66. Points on a Cylinder. In locating points on a cylindrical surface the elements of the cylinder must be utilized. Fig. 145a and b illustrates the principle involved in fixing the location of a given point P.

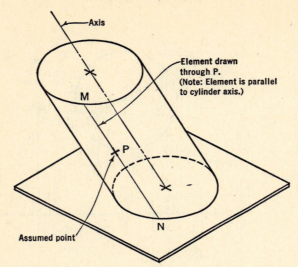

Fig. 145a

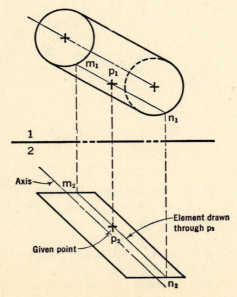

Fig. 145b

Example: Points on a cylinder. (Refer to Fig. 145.)

Given: Horizontal and vertical projections of an oblique cylinder and the vertical projection of point P (p_2).

Find: The horizontal projection (p_1) of point P.

Procedure:

    1. Draw element $m_2 n_2$ through p_2.

    2. Locate element $m_1 n_1$ by projection from view #2.

    3. Locate p_1 by projection from view #2. Point p_1 lies on $m_1 n_1$.

67. Piercing Points of a Line and a Cylinder. A cutting plane drawn *parallel to the axis* and *intersecting the base* of a cylinder cuts *straight-line elements* on the cylinder. If the cutting plane also contains a line which intersects the cylinder, the *piercing points* of the line will lie on the elements cut by the plane (see Fig. 146).

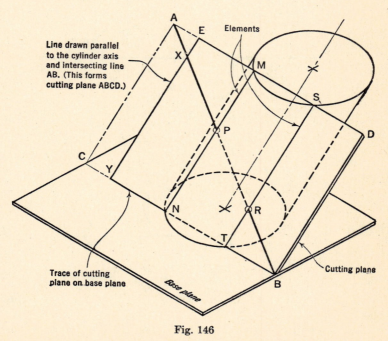

Fig. 146

Example: Piercing points of a line and a cylinder. (Refer to Fig. 147.)

Given: Horizontal and vertical projections of an oblique cylinder and intersecting line AB.

Find: The piercing points of line AB on the surface of the cylinder.

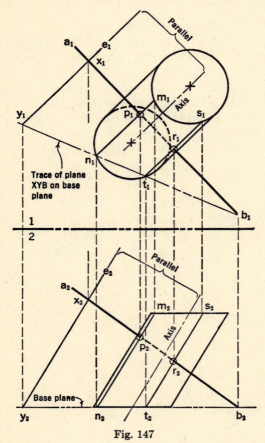

Fig. 147

Procedure:

1. Extend line AB in view #2 so that it intersects the base plane at b_2.

2. Draw any line (e_2y_2) **parallel** to the cylinder axis and intersecting a_2b_2 at x_2 and the base plane at y_2. (This creates a cutting plane which contains a_2b_2 and is parallel to the cylinder axis.)

3. Locate x_1 by projection from view #2.

4. Through x_1 draw a line parallel to the cylinder axis (in view #1). Determine y_1 by projection from view #2.

5. Connect y_1 and b_1 with a straight line. This is the trace of the cutting plane on the base plane. It cuts the base at n_1 and t_1.

6. Draw elements n_1m_1 and t_1s_1 (parallel to axis). The points at which these elements intersect a_1b_1 are the required piercing points (p_1 and r_1).

7. Locate p_2 and r_2 by projection. Determine visibility of line AB.

68. Tangent Planes to Cones and Cylinders. In determining the line of intersection between two intersecting surfaces, cutting planes are often used. Sometimes these cutting planes are *tangent* to one of the intersecting surfaces while cutting the other surface.

The following material will illustrate how tangent planes are determined orthographically.

69. Plane Tangent to a Cone at a Given Point on the Cone's Surface. Fig. 148 pictorially illustrates a plane tangent to a cone at a given point P on the cone's surface.

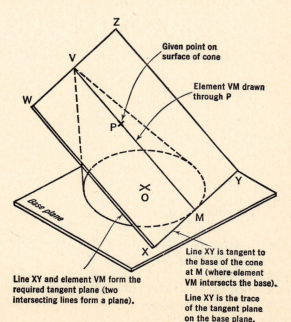

Given point on surface of cone

Element VM drawn through P

Fig. 148

Line XY and element VM form the required tangent plane (two intersecting lines form a plane).

Line XY is tangent to the base of the cone at M (where element VM intersects the base).

Line XY is the trace of the tangent plane on the base plane.

Example: Plane tangent to a cone at a given point on the surface of the cone. (Refer to Fig. 149.)

Given: Horizontal and vertical projections of cone VO, and the horizontal projection of point P on the surface of the cone.

Find: A plane tangent to cone VO at point P.

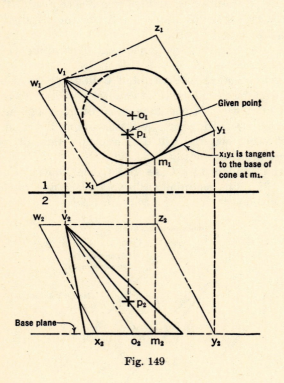

Fig. 149

Procedure:

   1. Draw element v_1m_1 through p_1.

   2. Find v_2m_2 and p_2 by projection.

   3. Draw line x_1y_1 *tangent* to the base of the cone at m_1.

   4. Lines VM and XY form plane $WXYZ$, which is the required tangent plane.

70. Plane Tangent to a Cone and Containing a Point Outside the Cone. Fig. 150 shows a plane tangent to a cone and containing a point which lies outside the cone's surface.

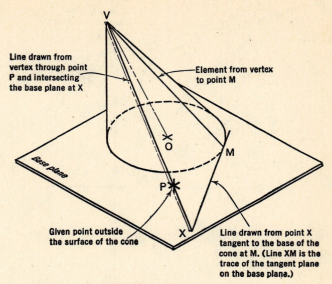

Line drawn from vertex through point P and intersecting the base plane at X

Element from vertex to point M

Base plane

O

M

P

Given point outside the surface of the cone

X

Line drawn from point X tangent to the base of the cone at M. (Line XM is the trace of the tangent plane on the base plane.)

Lines VX and XM form a plane tangent to the cone and containing point P.

Fig. 150

Example: Plane tangent to a cone and containing a given point outside the cone. (Refer to Fig. 151, p. 188.)

Given: Horizontal and vertical projections of cone *VO*, and point *P*.

Find: Tangent plane containing point *P*.

Procedure:

1. Draw a straight line from v_2 to p_2 and intersecting the base plane at x_2.

2. Draw a straight line from v_1 to p_1. Find x_1 by projection from view #2.

3. From x_1 draw a line tangent to the base of the cone (tangent point m_1).

4. Lines *VX* and *XM* form the required tangent plane *VXM*.

Note: A second tangent plane can be determined by drawing a line from x_1 tangent to another point on the cone's base.

71. Plane Tangent to a Cone and Parallel to a Line Outside the Cone. Fig. 152 (p. 188) pictorially illustrates a plane that is *tangent* to a cone and also *parallel* to a line which lies outside the cone.

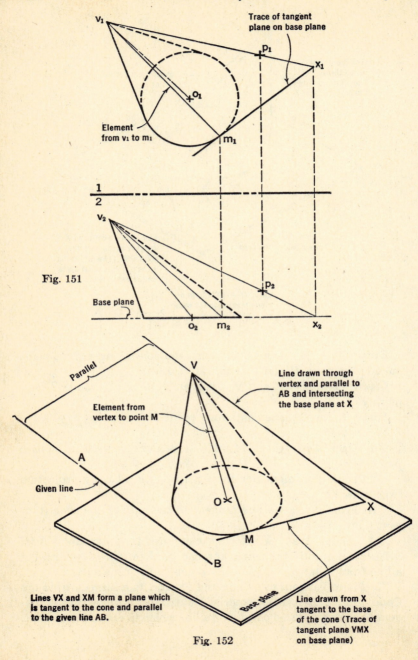

Fig. 151

Trace of tangent plane on base plane

Element from v_1 to m_1

Base plane

Parallel

Element from vertex to point M

Given line

Line drawn through vertex and parallel to AB and intersecting the base plane at X

Line drawn from X tangent to the base of the cone (Trace of tangent plane VMX on base plane)

Lines VX and XM form a plane which is tangent to the cone and parallel to the given line AB.

Fig. 152

Example: Plane tangent to a cone and parallel to a given line. (Refer to Fig. 153.)

Given: Horizontal and vertical projections of cone VO, and line AB.

Find: Tangent plane parallel to line AB.

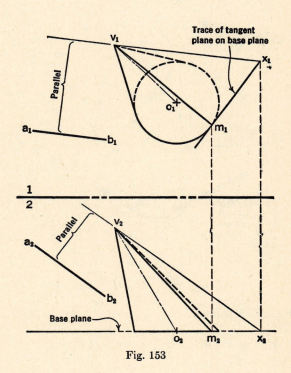

Fig. 153

Procedure:

1. In view #2 draw a line from v_2 parallel to a_2b_2 and intersecting the base plane at x_2.

2. In view #1 draw a line from v_1 parallel to a_1b_1. Determine x_1 by projection from view #2.

3. From x_1 draw a line *tangent* to the base of the cone (tangent point m_1).

4. Lines VX and XM determine a plane that is tangent to the cone and parallel to line AB.

72. Plane Tangent to a Cylinder at a Given Point on the Cylinder. Fig. 154a shows a plane tangent to a cylinder at a given point *P*.

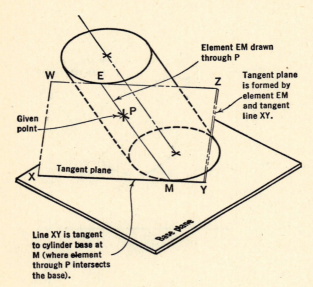

Element EM drawn through P

Tangent plane is formed by element EM and tangent line XY.

Given point

Tangent plane

Line XY is tangent to cylinder base at M (where element through P intersects the base).

Base plane

Fig. 154a

Example: Plane tangent to a cylinder at a given point. (Refer to Fig. 154*b*.)

Given: Horizontal and vertical projections of an oblique cylinder, and point *P*.

Find: Tangent plane at point *P*.

Procedure:

    1. Draw element e_1m_1 through p_1 and find e_2m_2 by projection.

    2. Draw x_1y_1 tangent to the cylinder base at m_1.

    3. Element *EM* and line *XY* form the required tangent plane (*WXYZ*).

73. Plane Tangent to a Cylinder and Containing a Point Outside the Cylinder. Fig. 155 pictorially illustrates a plane tangent to a cylinder and containing a point which lies outside the surface of the cylinder.

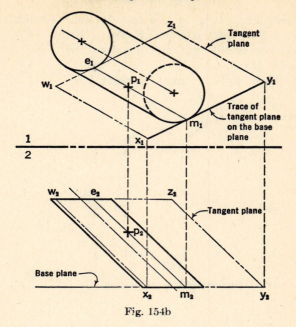

Fig. 154b

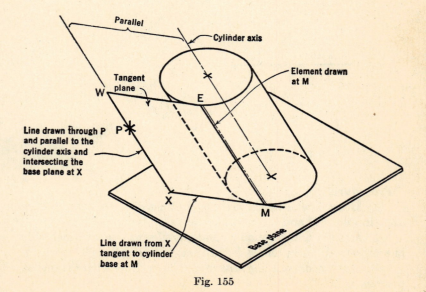

Fig. 155

Example: Plane tangent to a cylinder and containing a given point outside the surface of the cylinder. (Refer to Fig. 156.)

Given: Horizontal and vertical projections of an oblique cylinder, and point P.

Find: Tangent plane containing point P.

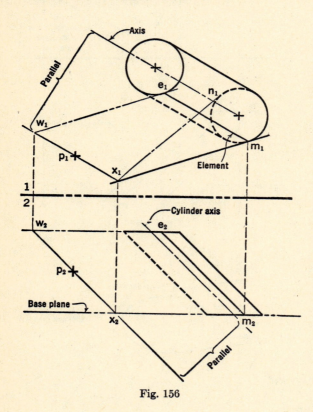

Fig. 156

Procedure:

1. Draw line w_2x_2 through p_2 and *parallel* to the cylinder axis (in view #2) and intersecting the base plane at x_2.

2. Draw a line through p_1 and *parallel* to the cylinder axis in view #1. Find w_1 and x_1 by projection from view #2.

3. From x_1 draw a line *tangent* to the cylinder's base (tangent point m_1).

4. At m_1 draw element e_1m_1. Element *EM* and line *XM* form the required tangent plane.

5. Find the tangent plane in view #2 by projection.

Note: A second tangent plane can be formed by drawing a line from x_1 tangent to another point (n_1) on the cylinder's base.

74. Plane Tangent to a Cylinder and Parallel to a Line Outside the Cylinder. Fig. 157 shows a plane which is *tangent* to a cylinder and also *parallel* to a line which lies outside the surface of the cylinder.

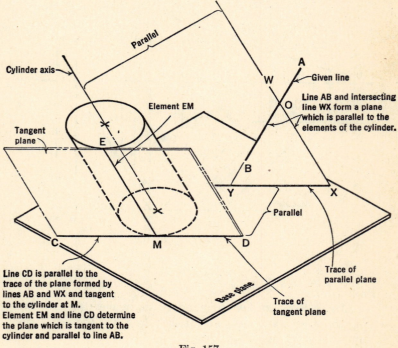

Fig. 157

Example: Plane tangent to a cylinder and parallel to a given line outside the surface of the cylinder. (Refer to Fig. 158.)

Given: Horizontal and vertical projections of an oblique cylinder, and line *AB*.

Find: Tangent plane parallel to line *AB*.

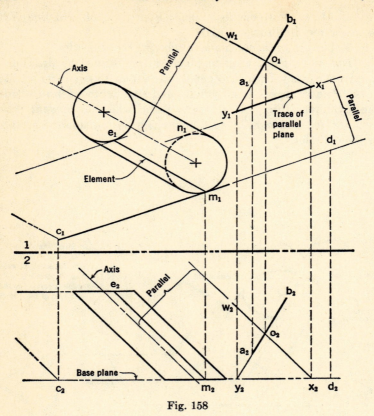

Fig. 158

Procedure:

1. Draw w_2x_2 **parallel** to cylinder axis in view #2 and intersecting the base plane at x_2 and intersecting b_2y_2 (a_2b_2 extended) at any convenient point o_2.

2. Find o_1 by projection. At o_1 draw w_1x_1 parallel to cylinder axis in view #1. Locate x_1y_1 (trace of parallel plane) by projection from view #2.

3. Draw c_1d_1 **parallel** to x_1y_1 and **tangent** to cylinder base (tangent point m_1). At m_1 draw element e_1m_1.

4. Element EM and line CD determine the required tangent plane.

5. Find projection of tangent plane in view #2.

Note: A second tangent plane can be determined by drawing a line parallel to x_1y_1 at point n_1. These two planes at m_1 and n_1 are the limiting planes. Between these two planes any number of planes can be drawn which are either tangent or secant to the base of the cylinder and at the same time are parallel to the line outside the cylinder.

Practice Problems

VIII-1. Given: Horizontal and vertical projections of a triangular oblique prism and the horizontal projection of a point P on one face of the prism.

Find: The vertical projection of point P.

VIII-2. Given: Horizontal and vertical projections of an oblique prism and an intersecting line LM.

Find: (a) The piercing points of the line LM and the given prism.

(b) Complete the visibility of line LM.

VIII-3. Given: Horizontal and vertical projections of a triangular prism and an intersecting line XY.

Find: (a) The piercing points of line XY and the given prism.

(b) Complete the visibility of line XY.

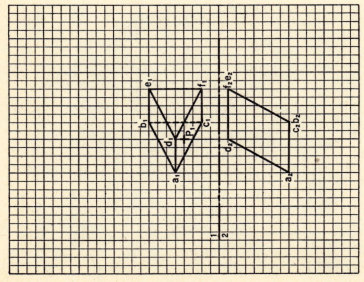

VIII-1

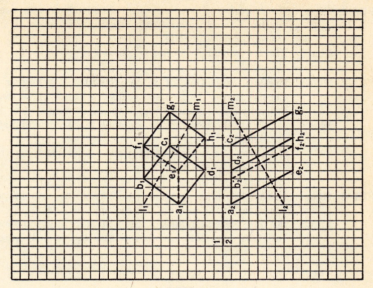

VIII–2

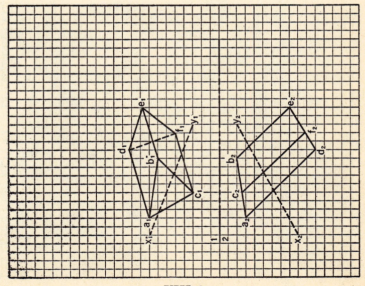

VIII–3

VIII-4. Given: Horizontal and vertical projections of a right cone (vertex V) and the vertical projection of a point P on the surface of the cone.

        Find: The horizontal projection of point P.

VIII-5. Given: Vertical projection of a cone (vertex V) having a circular base 2″ in diameter (center A) and the vertical projection of a point Q on the cone's surface.

        Find: (a) Draw the top view of the cone and show complete visibility.

               (b) Draw the horizontal projection of point Q on the surface of the given cone.

VIII-6. Given: Horizontal and vertical projections of an oblique cone (vertex V) and an intersecting line LM.

        Find: (a) The piercing points of line LM and the given cone.

               (b) Show complete visibility of line LM in both given views.

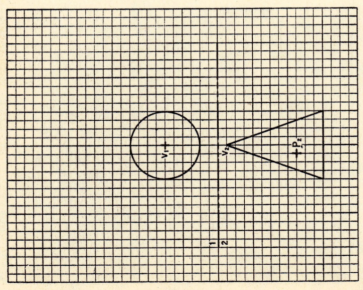

VIII–4

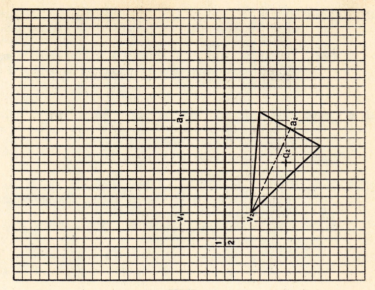

VIII–5

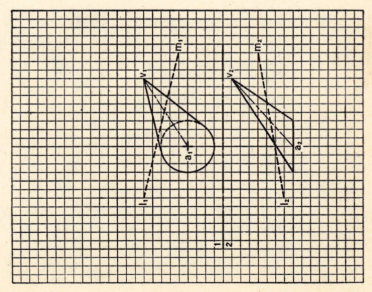

VIII–6

VIII–7. Given: Horizontal and vertical projections of an oblique cone (vertex V and base parallel to the vertical projection plane) and an intersecting line XY.

Find: (a) The piercing points of line XY and the given cone.

(b) Show complete visibility of line XY in both given views.

VIII–8. Given: Horizontal and vertical projections of an oblique cylinder (axis OP) and the horizontal projection of a point S on the surface of the cylinder.

Find: The vertical projection of point S.

VIII–9. Given: Horizontal and vertical projections of an oblique cylinder (axis AB parallel to the profile projection plane) and an intersecting line XY.

Find: (a) The piercing points of line XY and the given cylinder.

(b) Show complete visibility of XY in all views.

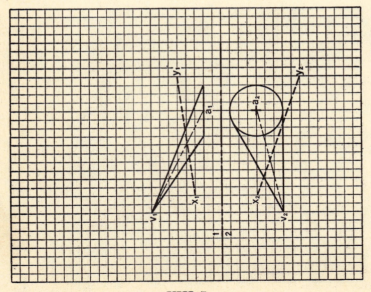

VIII–7

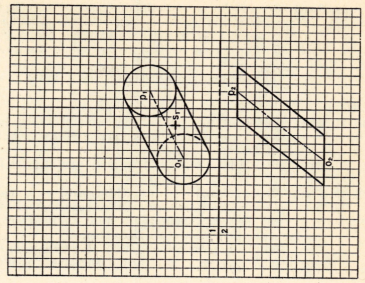

VIII–8

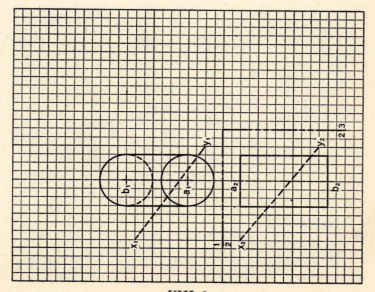

VIII–9

VIII–10. Given: Horizontal and vertical projections of an oblique cylinder (axis *AB* and base parallel to the vertical projection plane) and an intersecting line *LM*.

Find: (a) The piercing points of the line *LM* and the given cylinder.

(b) Show complete visibility of line *LM* in the given views.

VIII–11. Given: Horizontal and vertical projections of an oblique cone (vertex *V*) and the vertical projection of point *P* on the surface of the cone.

Find: Draw a plane tangent to the given cone at point *P*.

VIII–12. Given: Horizontal and vertical projections of an oblique cone (vertex *O*), the horizontal projection of point *Q*, and the vertical projection of point *P*. Both points are on the surface of the given cone.

Find: (a) Draw a plane tangent to the given cone at point *P*.

(b) Draw a plane tangent to the given cone at point *Q*.

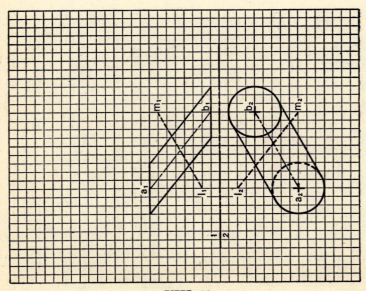

VIII–10

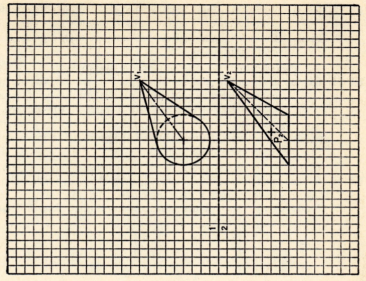

VIII–11

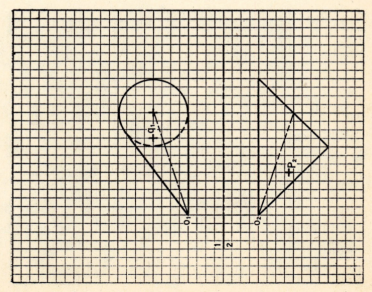

VIII–12

VIII–13. Given: Horizontal and vertical projections of an oblique cone (vertex V) and a point P outside the cone.

   Find: Draw two planes that are tangent to the given cone and contain point P.

VIII–14. Given: Horizontal and vertical projections of an oblique cone (vertex O and base parallel to the profile projection plane) and a point X outside the surface of the cone.

   Find: Draw two planes that are tangent to the given cone and contain the point X.

VIII–15. Given: Horizontal and vertical projections of an oblique cone (vertex V) and line LM.

   Find: Draw two planes that are parallel to line LM and tangent to the given cone.

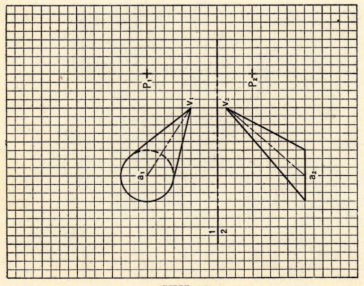

VIII–13

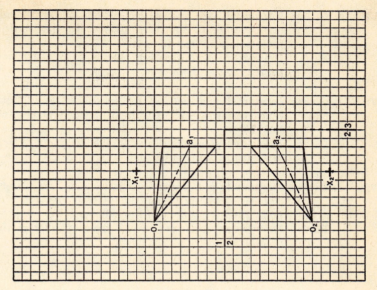

VIII–14

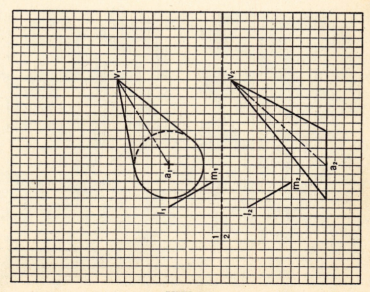

VIII–15

VIII–16. Given: Horizontal and vertical projections of an oblique cone (vertex *O* and base parallel to the vertical projection plane) and line *XY*.

Find: Draw two planes that are parallel to line *XY* and tangent to the given cone.

VIII–17. Given: Horizontal and vertical projections of an oblique cone (vertex *V* and base in an inclined plane) and line *LM*.

Find: Draw two planes that are parallel to line *LM* and tangent to the given cone.

VIII–18. Given: Horizontal and vertical projections of an oblique cylinder (axis *AB*) and the vertical projection of point *P* on the surface of the cylinder.

Find: Draw a plane tangent to the given cylinder at the point *P*.

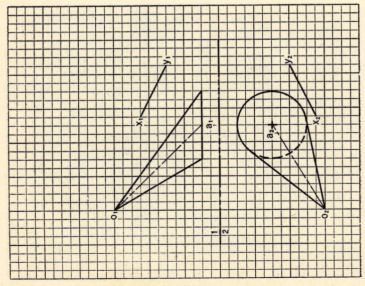

VIII–16

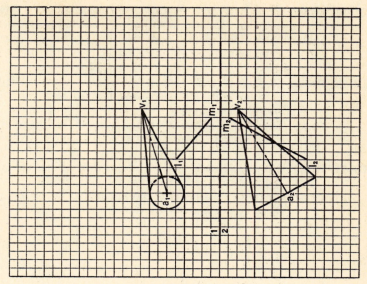

VIII–17

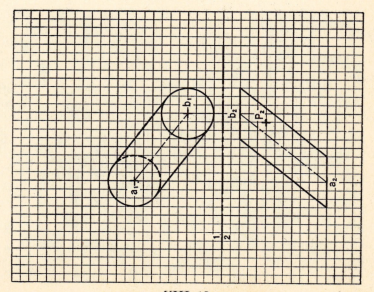

VIII–18

VIII-19. Given: Horizontal and vertical projections of an oblique cylinder (axis AB) and point P outside the cylinder.

Find: Draw two planes that are tangent to the given cylinder and contain the point P.

VIII-20. Given: Horizontal and vertical projections of an oblique cylinder (axis RS and bases frontal) and point X outside the cylinder.

Find: Draw two planes that are tangent to the given cylinder and contain point X.

VIII-21. Given: Horizontal and vertical projections of an oblique cylinder (axis AB) and line LM.

Find: Draw two planes that are parallel to line LM and tangent to the given cylinder.

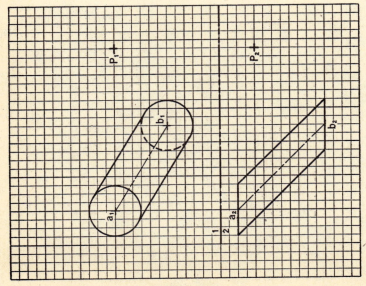

VIII–19

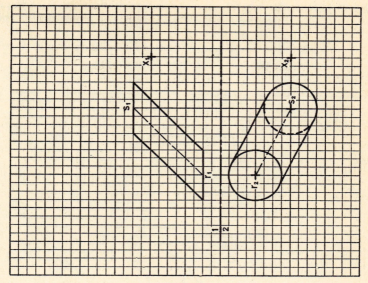

VIII–20

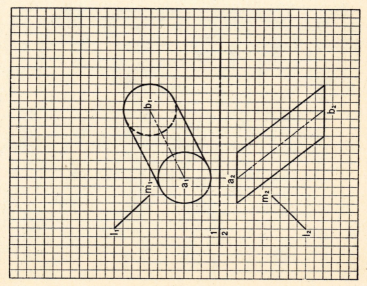

VIII–21

VIII–22. Given: Horizontal and vertical projections of an oblique cylinder (axis *AB* and base in a profile plane) and line *LM*.

Find: Draw two planes that are parallel to line *LM* and tangent to the given cylinder.

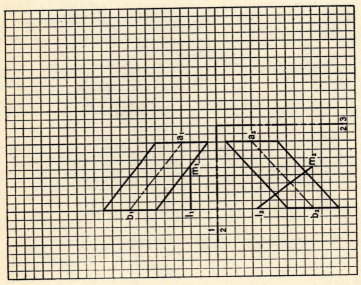

VIII–22

IX

Intersection of Surfaces

75. Descriptive Geometry Principles Used to Determine the Line of Intersection between Various Surfaces. This chapter will show how the principles which have been presented, in the chapters listed below, are applied to determine the lines of intersection between various surfaces.

76. Intersection of a Plane and a Prism (**Intersecting Plane Seen as an Edge**). A view where the intersecting plane appears as an edge will show where each *lateral edge* of the prism *pierces* the intersecting plane. These piercing points determine the line of intersection of the prism and plane. See Fig. 159.

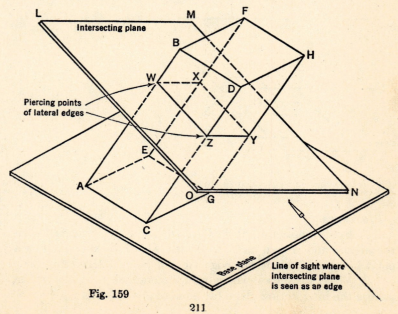

Fig. 159

Example: Intersection of a plane and a prism. (Refer to Fig. 160.)

Given: Horizontal and vertical projections of an oblique prism, and intersecting plane *LMNO*.

Find: Line of intersection between prism and plane *LMNO*.

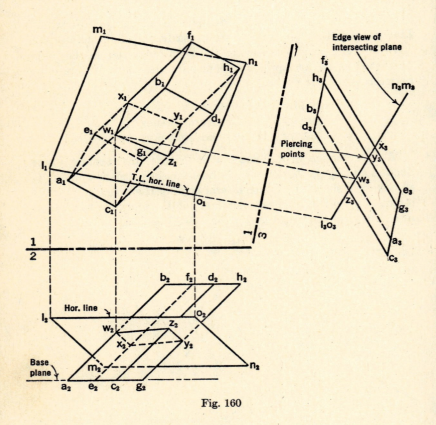

Fig. 160

Procedure:

1. Find intersecting plane *LMNO* as an edge. (*LO* is a horizontal line and is seen as T.L. in view #1. It is then seen as a point in view #3 where the edge view of plane *LMNO* appears.)

2. View #3 shows where each lateral edge of the prism pierces the intersecting plane (at points w_3, x_3, y_3, and z_3).

3. Project piercing points w_3, x_3, y_3, and z_3 to view #1. (Point w_1 lies on a_1b_1, point x_1 lies on e_1f_1, etc.)

4. Connect points w_1, x_1, y_1, and z_1 with straight lines. This is the horizontal projection of the required line of intersection. Determine visibility by inspection.

5. Find the vertical projection of the line of intersection by projection from view #1. Determine visibility by inspection.

77. Intersection of a Plane and a Prism (Vertical Cutting Plane Method). Fig. 161 shows how the principle of using vertical cutting planes (discussed in Secs. 37 and 63) is applied to determine the line of intersection between a plane and a prism.

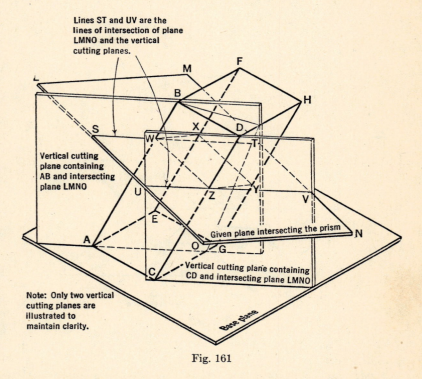

Fig. 161

Note: Only two vertical cutting planes are illustrated to maintain clarity.

Example: Intersection of a plane and a prism. (Refer to Fig. 162.)

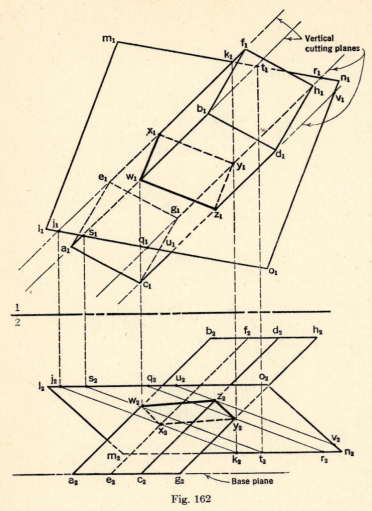

Fig. 162

Given: Horizontal and vertical projections of an oblique prism, and intersecting plane *LMNO*.

Find: Line of intersection between prism and plane *LMNO* using the vertical cutting plane method.

Procedure:

1. Draw a vertical cutting plane through lateral edge a_1b_1 (cutting plane appears as an edge in view #1). This plane intersects $l_1m_1n_1o_1$ at s_1 and t_1.

2. Find s_2 and t_2 by projection. Connect these points with a straight line. Where s_2t_2 intersects lateral edge a_2b_2 is the piercing point w_2 of the lateral edge on plane *LMNO*.

3. Find w_1 by projection.

4. Repeat the above process for each lateral edge.

5. Connect *WXYZ* in both views with straight lines. This is the required line of intersection.

6. Determine visibility.

78. Intersection of a Pyramid and a Prism. If the prism is viewed from a position where its *end view* can be seen, this view will show where the *edges* of the *pyramid* intersect the *faces* of the *prism*. The piercing points of the *edges* of the *prism* on the *pyramid faces* can be determined by passing cutting planes through each *prism edge* and intersecting the faces of the pyramid. See Fig. 163 and refer to Secs. 36 and 37.

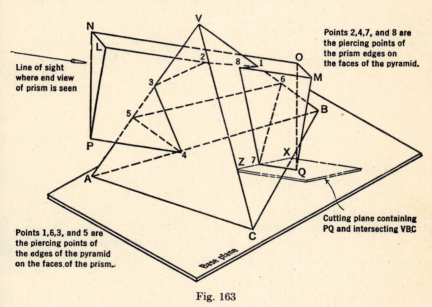

Line of sight where end view of prism is seen

Points 2,4,7, and 8 are the piercing points of the prism edges on the faces of the pyramid.

Cutting plane containing PQ and intersecting VBC

Points 1,6,3, and 5 are the piercing points of the edges of the pyramid on the faces of the prism.

Base plane.

Fig. 163

Example: Intersection of a pyramid and a prism. (Refer to Fig. 164.)

Given: Horizontal and vertical projections of a right triangular pyramid and an intersecting prism.

Find: The line of intersection of the pyramid and the prism.

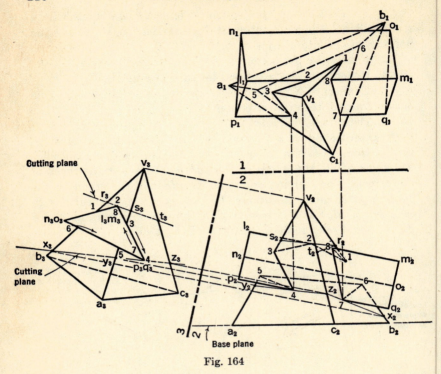

Fig. 164

Procedure:

1. Draw the **end view** of prism $LMNOPQ$ (see view #3).

2. In view #3 number consecutively each piercing point, starting with points on the **visible** faces of the pyramid and continuing to the piercing points on the **hidden** face of the pyramid. For example: points 1, 2, 3, 4, 5, and 6 are **visible** points in view #3; points 7 and 8 lie on face $v_3b_3c_3$ — which is a **hidden** face in view #3.

3. Locate points 2, 4, 7, and 8 in views #2 and #1 by passing cutting planes through the prism edges on which these points lie and intersecting the faces of the pyramid (see view #3).

4. Project all piercing points to views #2 and #1.

5. Connect all the points in consecutive order (views #2 and #1). Determine the visibility of the line of intersection by determining the visibility of the **individual** piercing points. For ex-

ample: to be visible a point must lie on a *visible* edge or face of the prism *and* on a *visible* edge or face of the pyramid.

6. Line 1–2–3–4–5–6–7–8 is the required line of intersection.

79. Intersection of Two Cylinders (Oblique — Bases in One Plane). By passing cutting planes *parallel* to the axes of two *intersecting* cylinders, *elements* are cut from both cylinders. When one element on one cylinder intersects an element on the other cylinder (both cut by the same plane) a point *common* to *both* cylinders is located. This point lies on the line of intersection of the two cylinders. By passing a number of cutting planes through the cylinders the line of intersection can be determined. See Fig. 165a and 165b.

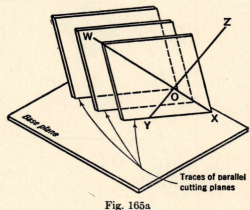

Fig. 165a

Example: Intersection of two cylinders. (Refer to Fig. 167, pp. 220–221.)

Given: Horizontal and vertical projections of two oblique intersecting cylinders having bases in one plane.

Find: The line of intersection of the two cylinders.

Procedure:

1. Create a plane which is *parallel* to axes AB and CD, and locate the *trace* (x_1y_1) of this plane on the *base plane*. Lines drawn parallel to this trace (in the base plane) represent *traces* of planes which are parallel to both cylinder axes.

2. Draw a convenient number of cutting plane traces (parallel to x_1y_1) through *both* cylinder bases. Cutting planes *must* cut elements from *both cylinders*. The *limiting* cutting planes are *tangent* to one cylinder and *secant* to the second cylinder. (Cut-

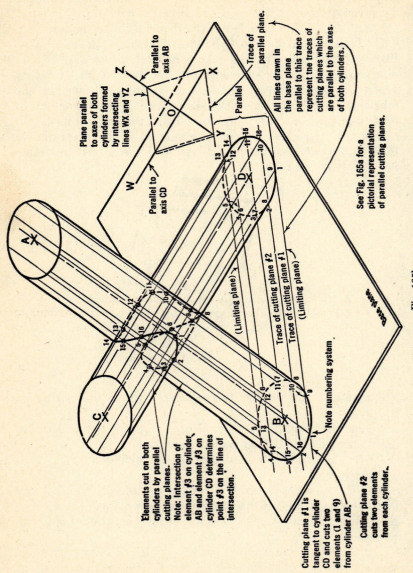

Plane parallel to axes of both cylinders formed by intersecting lines WX and YZ

Parallel to axis AB

Parallel to axis CD

Trace of parallel plane.

All lines drawn in the base plane parallel to this trace represent the traces of cutting planes which are parallel to the axes. of both cylinders.)

Parallel

(Limiting plane)

Trace of cutting plane #2

Trace of cutting plane #1

(Limiting plane)

Note numbering system.

See Fig. 165a for a pictorial representation of parallel cutting planes.

Base plane

Elements cut on both cylinders by parallel cutting planes. Note: Intersection of element #3 on cylinder AB and element #3 on cylinder CD determines point #3 on the line of intersection.

Cutting plane #1 is tangent to cylinder CD and cuts two elements (1 and 9) from cylinder AB.

Cutting plane #2 cuts two elements from each cylinder.

Fig. 165b

ting plane traces drawn beyond the limiting planes cut elements from only one cylinder and therefore cannot locate points common to both cylinders.)

3. Apply the numbering system illustrated in **Fig. 166**. Start numbering at a secant point. (Note only one number at secant points determined by the tangent planes.)

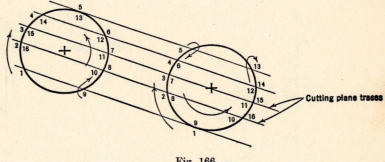

Fig. 166

4. Draw the elements cut by the cutting planes.

5. Where *like* number elements intersect, a point common to both cylinders is located. To be visible a point must be determined by *two* visible elements.

6. Connect points in *numerical order* with a smooth curve. This is the required line of intersection.

7. Determine the line of intersection in view #2 by projection from view #1.

8. Determine the visibility of the curve in both views.

80. Intersection of Two Cones (Oblique — Bases in One Plane). Cutting planes which pass through the *vertices* of two intersecting cones cut *straight-line elements* from both cones. A line (vertex line) connecting the vertices of both cones is contained by all cutting planes which cut elements from both cones. All cutting planes therefore *hinge* on this *vertex line*. When one element on one cone intersects an element of the other cone (both cut by the same plane) a point common to both cones is located. By passing a number of cutting planes through the two cones (and the vertex line) the line of intersection can be determined. See **Fig. 168** and **Fig. 170**.

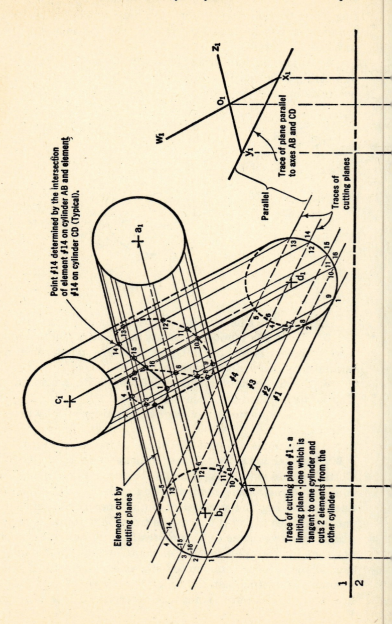

Point #14 determined by the intersection of element #14 on cylinder AB and element #14 on cylinder CD (Typical).

Trace of plane parallel to axes AB and CD

Parallel

Traces of cutting planes

Elements cut by cutting planes

Trace of cutting plane #1 - a limiting plane - one which is tangent to one cylinder and cuts 2 elements from the other cylinder

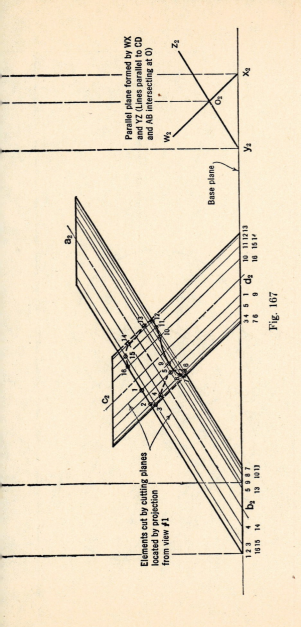

Fig. 167

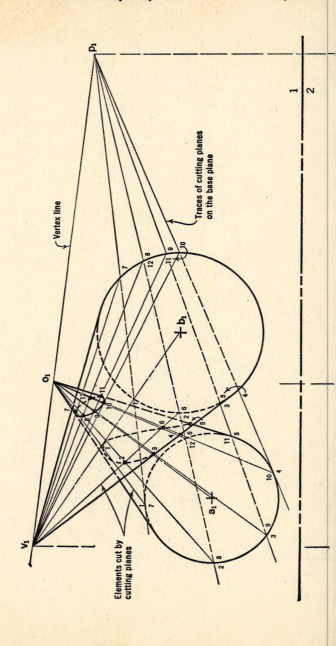

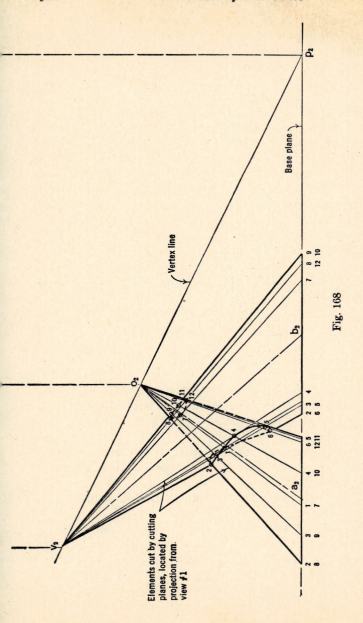

Fig. 168

Example: Intersection of two cones. (Refer to Fig. 168.)

Given: Horizontal and vertical projections of two oblique intersecting cones having bases in one plane.

Find: The line of intersection of the two cones.

Procedure:

1. Draw a line connecting vertices v_2 and o_2, and intersecting the *base plane* at p_2. This line is the *vertex line.*

2. Locate p_1 by projection from view #2. All cutting planes will hinge on the vertex line. Lines drawn through p_1 and through the bases of *both cones* represent *traces* of cutting planes on the *base plane* which cut straight-line elements from both cones.

3. Draw a convenient number of cutting plane traces which cut elements from both cones. (The limiting planes are tangent to cone *AO*.) Apply the numbering system illustrated in Fig. 169. Start numbering at a tangent point.

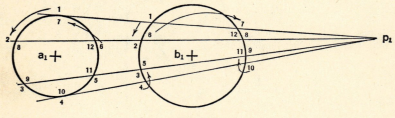

Fig. 169

4. Draw the elements cut by the cutting planes.

5. Where *like* number elements intersect, a point common to both cones is located. To be visible a point must be determined by *two* visible elements.

6. Connect points in numerical order with a smooth curve. Note that there are *two* lines of intersection in this example. These are the required lines of intersection.

7. Determine the lines of intersection in view #2 by projection from view #1.

81. Intersection of Two Cones (Bases in Different Planes).
To determine the line of intersection of two cones having bases in different planes, the *traces* of the cutting planes (which are hinged on the *vertex line* and cut straight-line elements from both cones) must be determined on *both base planes.* These traces indicate where each cutting plane cuts through the *base* of each cone. See Fig. 171 and Fig. 171a.

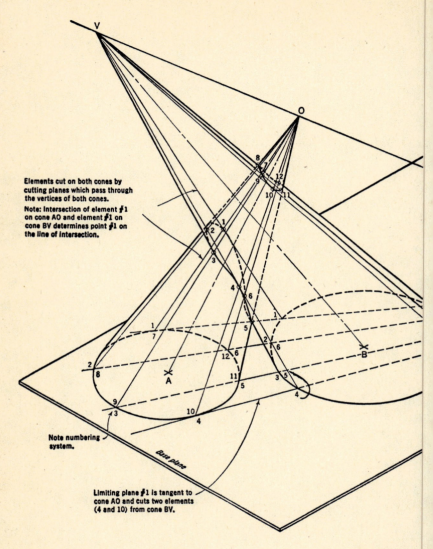

Elements cut on both cones by
cutting planes which pass through
the vertices of both cones.

Note: Intersection of element #1
on cone AO and element #1 on
cone BV determines point #1 on
the line of intersection.

Note numbering
system.

Base plane

Limiting plane #1 is tangent to
cone AO and cuts two elements
(4 and 10) from cone BV.

Fig. 170

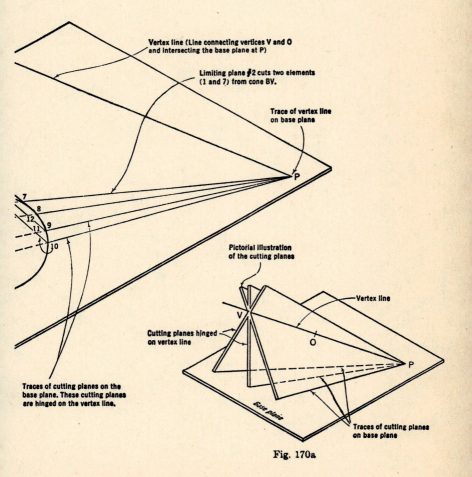

Vertex line (Line connecting vertices V and O and intersecting the base plane at P)

Limiting plane #2 cuts two elements (1 and 7) from cone BV.

Trace of vertex line on base plane

P

7
8
12
9
11
10

Pictorial illustration of the cutting planes

Vertex line

Cutting planes hinged on vertex line

V

O

P

Traces of cutting planes on the base plane. These cutting planes are hinged on the vertex line.

Base plane

Traces of cutting planes on base plane

Fig. 170a

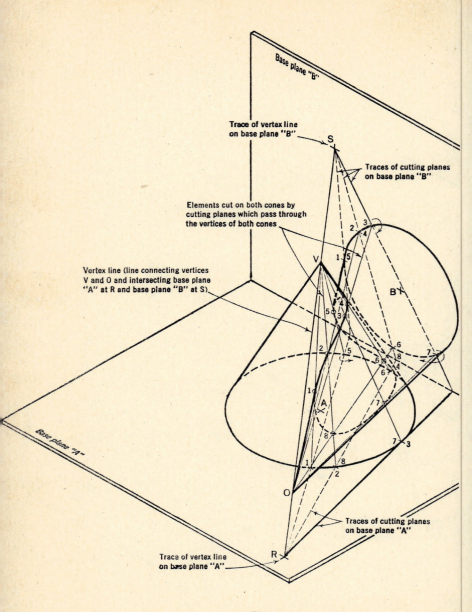

Base plane "B"

Trace of vertex line
on base plane "B"

S

Traces of cutting planes
on base plane "B"

Elements cut on both cones by
cutting planes which pass through
the vertices of both cones

Vertex line (line connecting vertices
V and O and intersecting base plane
"A" at R and base plane "B" at S)

B

V

A

O

Traces of cutting planes
on base plane "A"

Trace of vertex line
on base plane "A"

R

Base plane "a"

Fig. 171

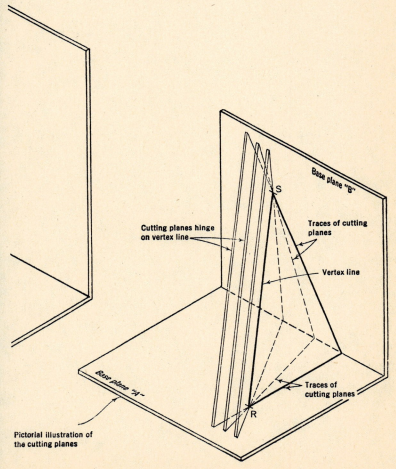

Base plane "B"

Cutting planes hinge
on vertex line

Traces of cutting
planes

Vertex line

S

Base plane "A"

Traces of
cutting planes

R

Pictorial illustration of
the cutting planes

Fig. 171a

Fig. 171 and 171a pictorially illustrate the intersection of two cones with bases in planes (*A* and *B*) which are perpendicular to each other. (The principles and procedures covered in this article can be applied to the intersection of two cones having bases in planes making any angle with each other.)

In Fig. 171 the line through the vertices *O* and *V* intersects the base planes of the intersecting cones at *R* and *S* respectively. When a cutting plane is passed through the line *RS*, this plane (when extended) cuts each base once. (Note the traces of the cutting planes.) The traces of the cutting planes on the base planes *A* and *B* intersect on the line of intersection of the base planes (see Fig. 171a). If a cutting plane is tangent or secant to both cones, it determines an element on each cone. The point at which these two elements intersect is a point common to both cones. Note in Fig. 171 that the plane which is tangent to cone *V* at point #3 (at the base of the cone) is secant to cone *O* at points #3 and #7 (at the base of cone *O*). The element *V*–3 on cone *V* intersects the elements *O*–3 and *O*-7 on cone *O*. These two points of intersection are common to both cones and therefore lie on the line of intersection between the two cones.

In order that elements be determined by cutting planes which intersect at points common to both cones the following conditions must be true:

1. A cutting plane must be secant to both cones.
2. A cutting plane must be tangent to both cones.
3. A cutting plane must be tangent to one cone and secant to the other cone.

If a cutting plane is tangent to one cone but does not touch or cut the other it can be seen that only one element will be determined by the cutting plane on the cone to which it is tangent.

The orthographic example which follows illustrates how the principles of descriptive geometry are used to determine the line of intersection between two cones having their bases in different planes. In the presentation of this example, the development of the reasoning and method is broken down into a series of figures and steps which show how each point of the line of intersection is found.

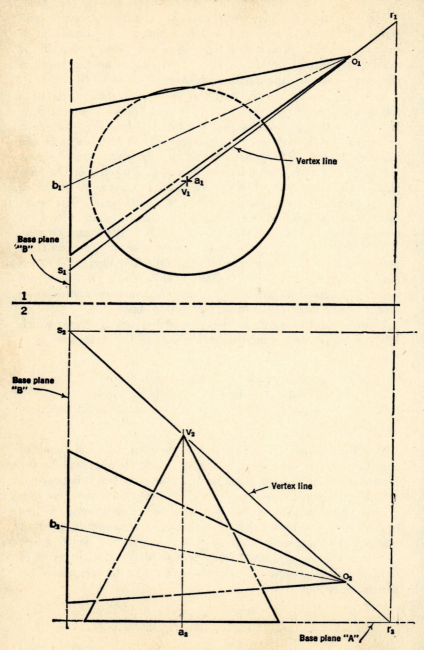

Fig. 172

Example: Line of intersection of two cones having bases in different planes. (Refer to Figs. 172, 172a, 172b, and 172c on pp. 226–233.)

Given: Horizontal, vertical, and profile projections of two cones with bases in different planes.

Find: The line of intersection of the two cones.

Procedure:

See Fig. 172

1. Draw the ***vertex line*** through vertices V and O of both given cones in all views.

2. Find where the vertex line pierces the Base Plane "A" and the Base Plane "B" (at points R and S respectively) in all views.

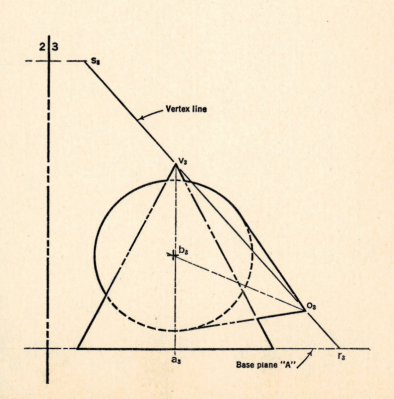

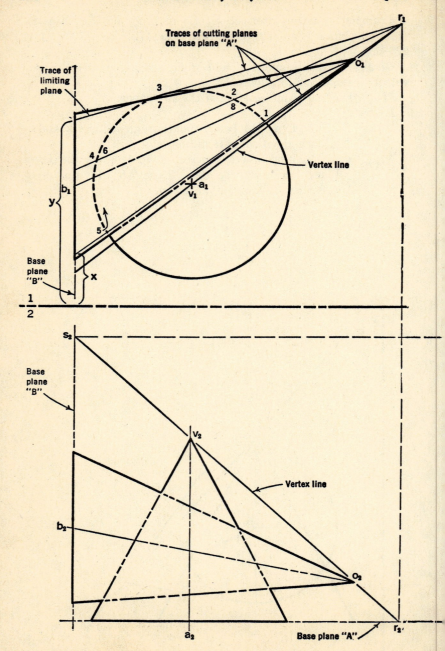

Fig. 172a

3. Draw a convenient number of cutting planes (indicated by their traces) in Base Plane "A" (view #1). Locate the profile projections of the traces of these cutting planes on the Base Plane "B" (view #3) by projection from view #1 (note distances X and Y). Note that one limiting plane is tangent to the base of cone VA in view #1 and that the other limiting plane is tangent to the base of cone OB in the profile view (#3).

See Fig. 172a

Only three cutting planes are used in the example to maintain clarity. In actual problem solution at least six cutting planes would be necessary to determine the line of intersection accurately.

4. Apply numbering system.

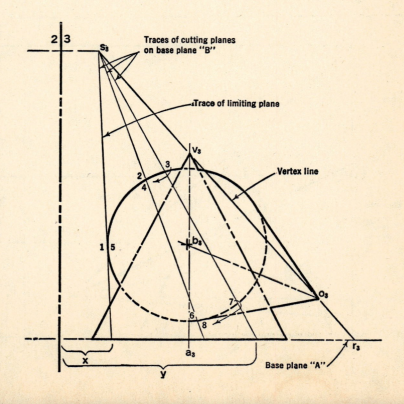

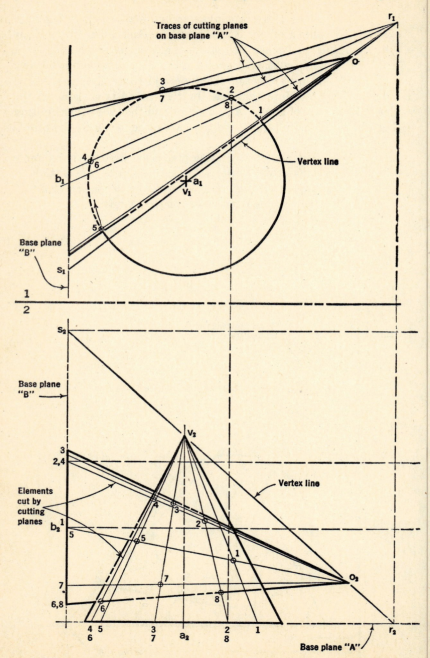

Traces of cutting planes
on base plane "A"

Vertex line

Base plane
"B"

Elements
cut by
cutting
planes

Vertex line

Base plane
"B"

Base plane
"A"

Fig. 172b

See
Fig. 172b

5. In the front view (#2) draw the elements cut on both cones by the cutting planes. Where *like* number elements intersect, points are determined on the line of intersection of the two cones.

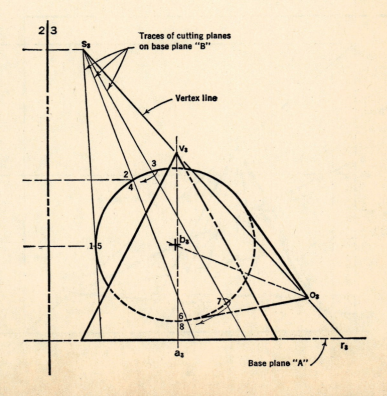

Traces of cutting planes on base plane "B"

Vertex line

Base plane "A"

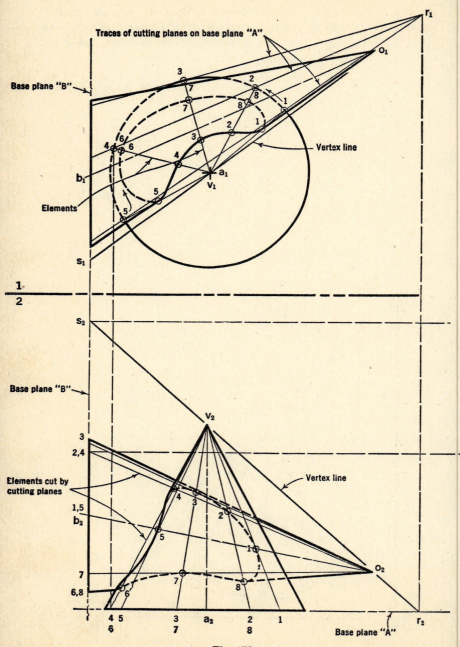

Traces of cutting planes on base plane "A"

Base plane "B"

Vertex line

Elements

Base plane "B"

Elements cut by
cutting planes

Vertex line

Base plane "A"

Fig. 172e

6. Connect all points located in view #2 with a smooth curve. This is the front view of the line of intersection.

7. Determine the visibility of the curve in the front view. To be visible a point must lie on a *visible element* of cone *VA* and on a *visible element* of cone *OB*.

See Fig. 172c for complete solution

8. Find the line of intersection in views #1 and #3 by drawing the elements cut on cone *VA* in view #1 and on cone *OB* in view #3. Project the points determined in view #2 to the proper elements in views #1 and #3.

9. Determine the visibility of the curve of intersection in views #1 and #3.

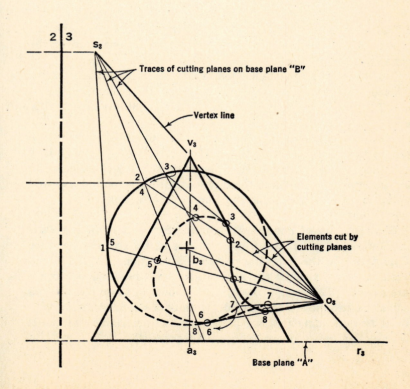

82. Intersection of a Cone and a Cylinder (Bases in One Plane).
Cutting planes which pass through the **vertex** of a cone and are
parallel to the **axis** of an intersecting cylinder cut straight-line
elements from the cone and the cylinder. These cutting planes
hinge on a line which is drawn through the **vertex** of the cone and
parallel to the cylinder **axis**. See Fig. 173.

Example: Line of intersection of a cone and a cylinder having
bases in the same plane. (Refer to Fig. 174.)

Given: Horizontal and vertical projections of a cone and an
intersecting cylinder.

Procedure:

1. Draw a line through v_2 and parallel to the cylinder axis
a_2b_2. (This is the front view of the hinge line.)

2. Find where the hinge line pierces the base plane (at point
P).

3. From p_1 in view #1 draw a convenient number of cutting
plane traces (only 5 used in example to maintain clarity).

4. Apply numbering system.

5. In view #1 draw the elements cut on the cone and the
cylinder by the cutting planes. Where **like** number elements intersect, points are determined on the line of intersection of the
cone and the cylinder.

6. Connect all points located in view #1 with a smooth curve.
This is the top view of the required line of intersection.

7. Determine the visibility of the line of intersection in view
#1.

8. Determine the front view of the line of intersection by projecting the elements cut by the cutting planes to view #2. Where
like number elements intersect, points on the line of intersection
are determined. Connect these points with a smooth curve and
determine the visibility of the line of intersection in the front
view (view #2).

**83. Summary of Possible Numbering Systems Involving Vari-
ous Intersections of Two Cones, a Cone and a Cylinder, or Two
Cylinders.** In the following pictorial illustrations (Figs. 175 and
175a, b, c, d), various types of intersections between *two cones*
are shown. The *numbering systems* shown in these illustrations
can also be applied to various intersections of a *cone and a cylinder,*
or of *two cylinders* (cutting planes used to determine the line of
intersection between two cylinders are parallel).

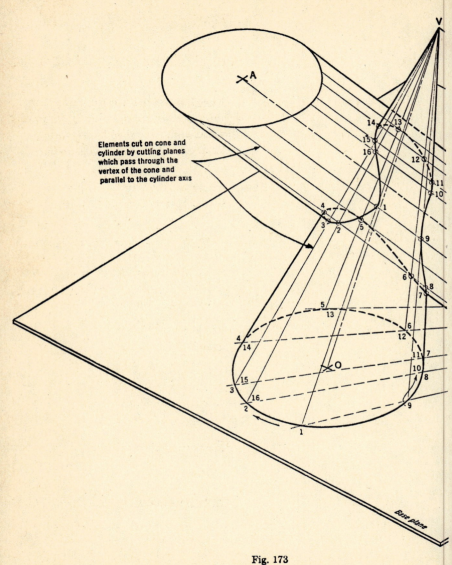

Elements cut on cone and cylinder by cutting planes which pass through the vertex of the cone and parallel to the cylinder axis

Fig. 173

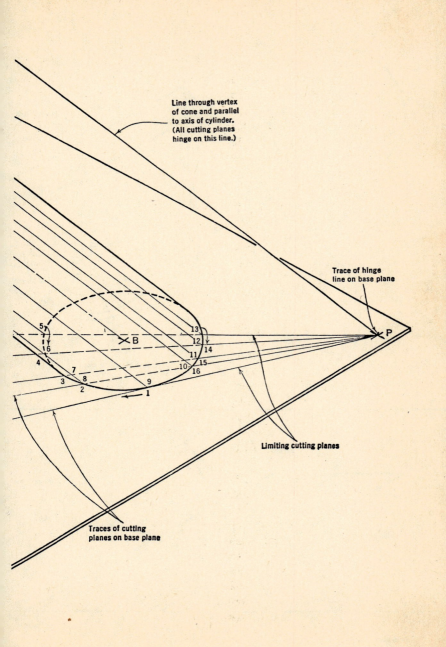

Line through vertex of cone and parallel to axis of cylinder. (All cutting planes hinge on this line.)

Trace of hinge line on base plane

Limiting cutting planes

Traces of cutting planes on base plane

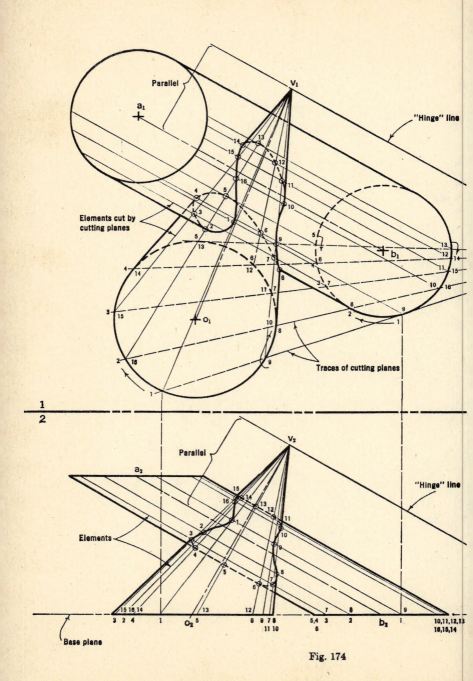

Fig. 174

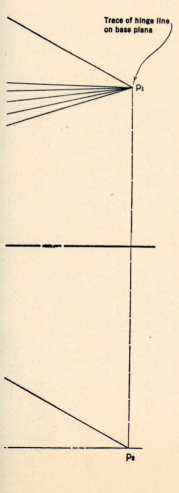

Trace of hinge line
on base plane

p₁

p₂

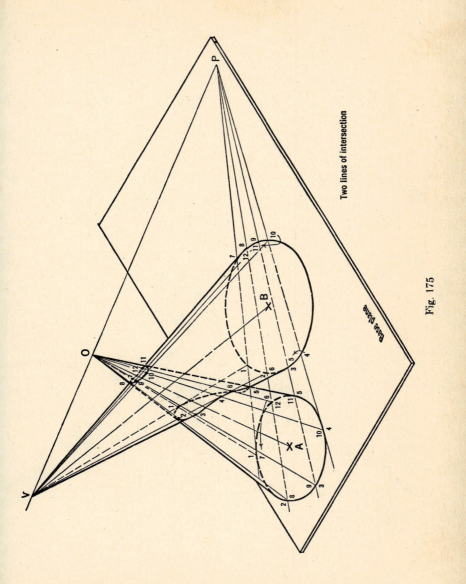

Fig. 175

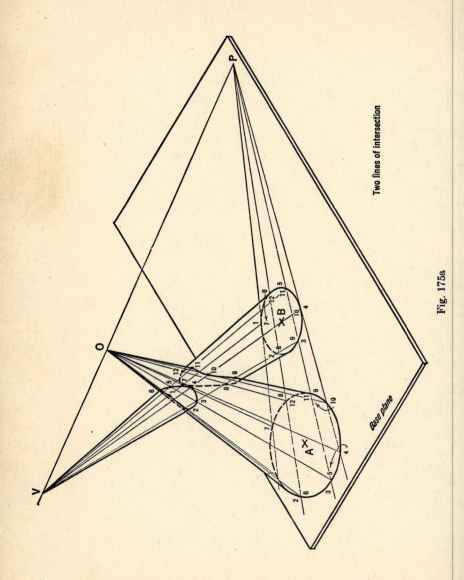

Fig. 175a

Base plane

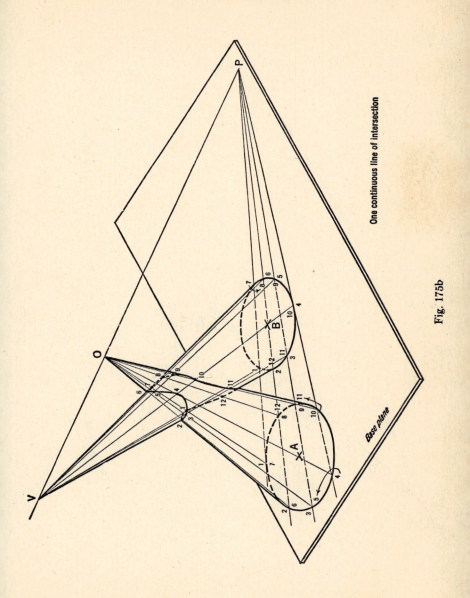

One continuous line of intersection

Base plane

Fig. 175b

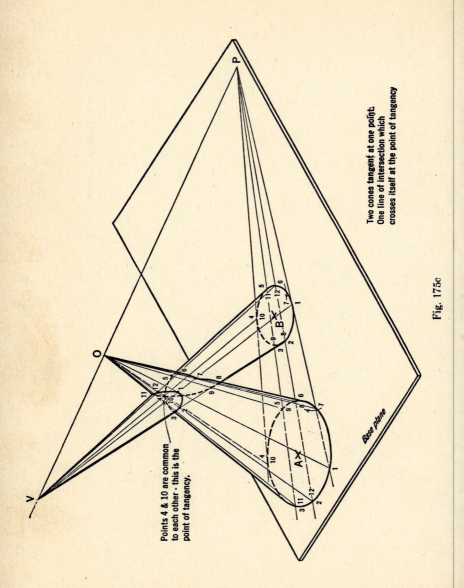

Points 4 & 10 are common
to each other - this is the
point of tangency.

Two cones tangent at one point;
One line of intersection which
crosses itself at the point of tangency

Fig. 175c

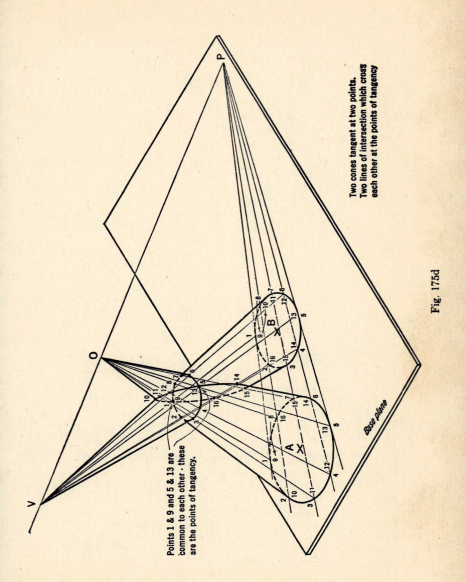

Points 1 & 9 and 5 & 13 are
common to each other - these
are the points of tangency.

Two cones tangent at two points.
Two lines of intersection which cross
each other at the points of tangency

Fig. 175d

Practice Problems

IX–1. Given: Horizontal and vertical projections of an oblique prism (bases ABC and EFG) and an intersecting plane WXY. (Both given objects shown in dotted lines.)

Find:
(a) The line of intersection of the given prism and the plane WXY by viewing WXY as an edge.

(b) Show visibility of the line of intersection and the two given figures.

(c) Reproduce the problem and solve by the vertical cutting plane method.

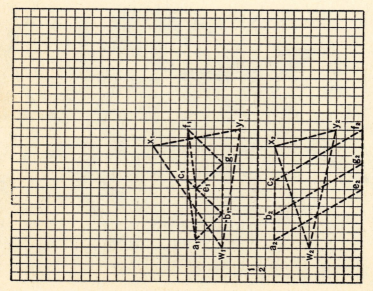

IX–1

IX-2. Given: Horizontal and vertical projections of an oblique triangular pyramid (vertex *V*, base *ABC*) and an intersecting plane *LMN*. (Both given figures shown in dotted lines.)

Find: (a) The line of intersection of the given pyramid and the plane *LMN* by viewing *LMN* as an edge.

(b) Show visibility of the line of intersection and the given figures in all views.

(c) Reproduce the problem and solve by the cutting plane method (see Sec. 37).

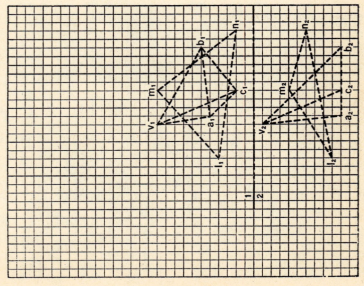

IX-2

IX–3. Given: Horizontal and vertical projections of an oblique cone (vertex V) and an intersecting plane WXY.

   Find: (a) The line of intersection of the given cone and the plane WXY by viewing WXY as an edge.

   (b) Show visibility of the line of intersection and the given figures in all views.

   Suggestion: Divide surface of cone into a number of elements and find where each element pierces the given plane. Connect the piercing points with a smooth curve.

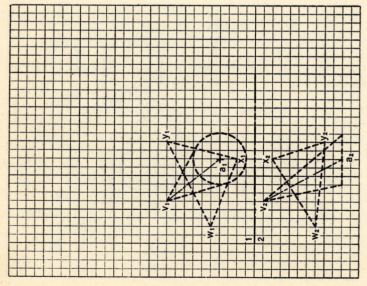

IX–3

IX–4. Given: Horizontal and vertical projections of an oblique triangular pyramid (vertex V, base ABC) and an intersecting triangular prism.

    Find: (a) The line of intersection of the pyramid and the given prism by viewing the prism from one end.

           (b) Show visibility of the line of intersection and both given figures in all views.

           (c) Reproduce the problem and find the line of intersection using only the given views. (Use cutting plane method.)

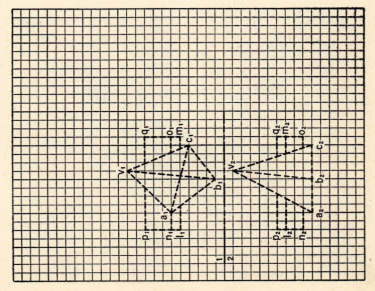

IX–4

IX-5. Given: Horizontal and vertical projections of an oblique cone (vertex *V*) and an intersecting triangular prism.

    Find: (a) The line of intersection between the prism and the given cone.

        (b) Show visibility of line of intersection and both figures in all views.

Suggestion: Divide surface of cone into a number of elements and handle problem as if the cone were a multisided pyramid.

IX-6. Given: Horizontal and vertical projections of two oblique intersecting cylinders (bases *A* and *B*).

    Find: (a) The line of intersection of the two cylinders.

        (b) Show the visibility of the line of intersection and the cylinders in both given views.

IX-7. Given: Horizontal and vertical projections of two oblique intersecting cylinders (bases *O* and *P* frontal).

    Find: (a) The line of intersection between the given cylinders.

        (b) Show the visibility of the line of intersection and the given figures in both given views.

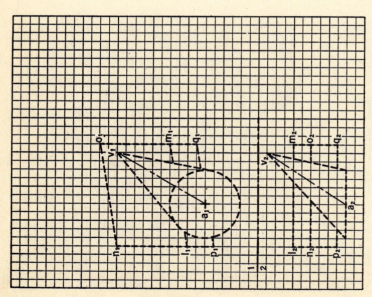

IX-5

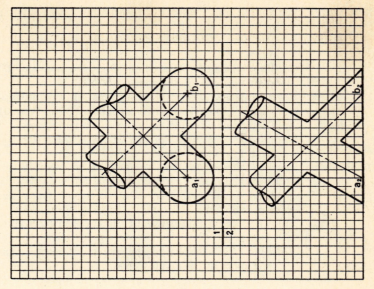

IX–6

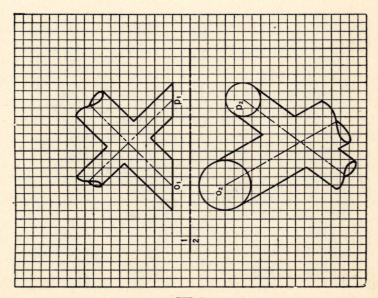

IX–7

IX–8. Given: Horizontal and vertical projections of two oblique cones (vertices *V* and *O*).

  Find: (a) The line of intersection between the two given cones.

  (b) Show visibility of the line of intersection and the given figures in both views.

IX–9. Given: Horizontal and vertical projections of two oblique intersecting cones (vertices *V* and *O* and bases frontal).

  Find: (a) The line of intersection between the two cones.

  (b) Show visibility of the line of intersection and the given figures in both views.

IX–10. Given: Horizontal and vertical projections of two oblique intersecting cones (vertices *V* and *O*) with circular bases of equal diameter. (Base *A* is horizontal and base *B* is profile.)

  Find: (a) The line of intersection of the cones in both given views.

  (b) Show the visibility of the line of intersection and the cones in the given views.

  Suggestion: Use two sheets of paper for this solution.

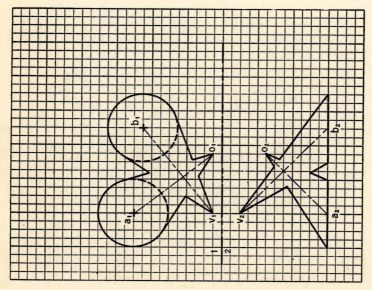

IX–8

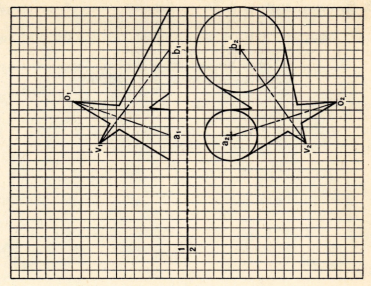

IX–9

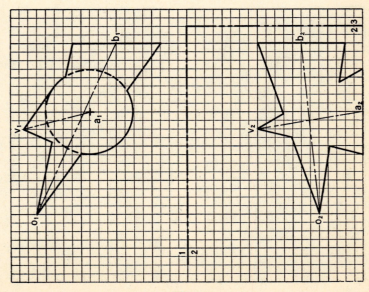

IX–10

IX-11. Given: Horizontal and vertical projections of an oblique cone and an oblique intersecting cylinder.

Find: (a) The line of intersection between the cone and cylinder in both given views.

(b) Show the visibility of the line of intersection and the given figures in both given views.

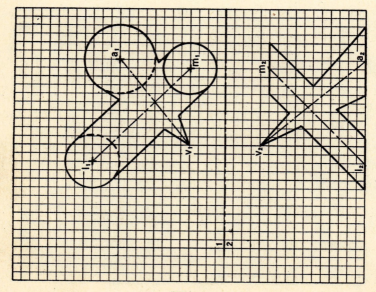

IX-11

IX-12. Given: Horizontal and vertical projections of an oblique cone (vertex V) and an intersecting cylinder (axis LM) both having frontal bases.

     Find: (a) The line of intersection between the given cone and cylinder in both views.

            (b) Show the visibility of the line of intersection and the given figures in the given views.

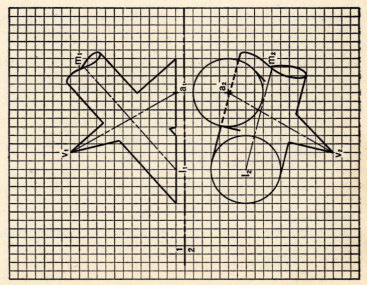

IX–12

X

Development of Surfaces

84. Definition of Surface Development. The surface or faces of an object normally can be represented orthographically by showing one or more views of the object. When the surface or faces of the object are *unrolled* or *laid out* onto a flat plane, the result is the *development* of the total surface of the given object. All lines in a development of an object appear *true length;* therefore all the faces or surfaces of the object appear in *true size.*

Certain surfaces such as double-curved and warped surfaces cannot be developed accurately (see Chapter VIII). However, *approximate developments* of these surfaces are possible. Double-curved surfaces can be approximately developed by assuming that these surfaces consist of small sections of developable surfaces such as cones and cylinders. Warped surfaces can be approximately developed by the method of *triangulation* which will be illustrated in this chapter.

85. Development of a Right Prism. Each face of the prism will appear *true size* in the development. (See Fig. 176a and Fig. 176b for a pictorial representation of the development.)

Example: Development of a right prism. (Refer to Fig. 177.)

Given: Horizontal and vertical projections of a right prism.

Find: Develop the lateral surface of the prism and attach the top and bottom bases to the development.

Note: Numbers instead of letters are used to denote the projections of the prism. This facilitates the drawing of the development.

Procedure:

1. Draw a line (stretch-out) upon which the actual *perimeter* of the given prism can be laid off.

2. With dividers lay off on the stretch-out line distances 1 to 2, 2 to 3, 3 to 4, and 4 to 1 (measured in view #1 where the *end view* of the prism is seen).

250

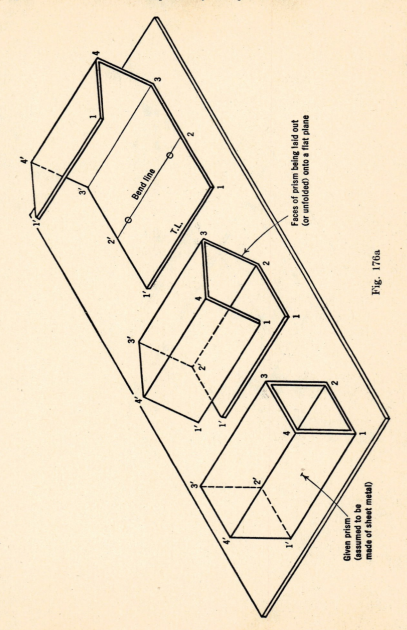

Bend line

T.L.

Faces of prism being laid out
(or unfolded) onto a flat plane

Given prism
(assumed to be
made of sheet metal)

Fig. 176a

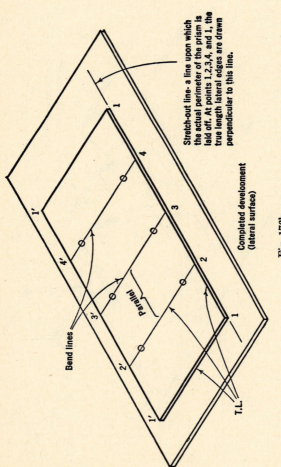

Stretch-out line- a line upon which the actual perimeter of the prism is laid off. At points 1,2,3,4, and 1, the true length lateral edges are drawn perpendicular to this line.

Bend lines

Parallel

T.L.

Completed development (lateral surface)

Fig. 176b

Note: Numbers instead of letters are used to denote the projections of the prism. This facilitates the drawing of the development.

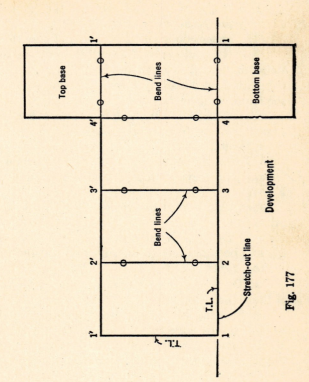

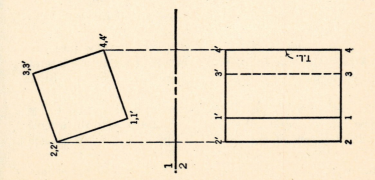

Fig. 177

3. At points 1, 2, 3, 4, and 1 on the stretch-out line draw the *true lengths* (1–1', 2–2', etc.) of the respective lateral edges of the prism *perpendicular* to the stretch-out line.

4. Connect points 1', 2', 3', 4', and 1' with straight lines.

5. Attach the true shape of the bottom base and the true shape of the top base to the lateral surface development of the prism (lines 4–1 and 4'–1' become bend lines). This completes the required development.

Note: This is a *parallel-line development.* Since the lateral edges of a prism are parallel, these edges are represented by *parallel lines* (bend lines) in the development.

86. Development of a Truncated Right Prism.

Example: Development of a truncated right prism. (Refer to Fig. 178.)

Given: Horizontal and vertical projections of a truncated right prism.

Find: Develop the lateral surface of the prism and attach the top and bottom bases to the surface development.

Procedure:

1. Draw a line (stretch-out) upon which the actual perimeter of the given prism can be laid off.

2. Starting with the *shortest lateral edge* (since it is at the shortest lateral edge that welding or joining takes place when the development is bent to shape) lay off distances 3 to 4, 4 to 1, 1 to 2, and 2 to 3 on the stretch-out line (these distances are seen in view #1).

3. At points 3, 4, 1, 2, and 3 draw the *true lengths* (3–3', 4–4', etc.) of the respective lateral edges *perpendicular* to the stretch-out line.

4. Connect points 3', 4', 1', 2', and 3' with straight lines.

5. Attach the true shape of the bottom base to the lateral surface development of the prism. (Line 1–2 becomes a bend line.)

6. Find the true shape of the top base (which is oblique) by projection (see views #3 and #4). Attach this base to the lateral surface development of the prism. In this case the top base is attached to the *longest* side (1'–2'). This would be the most economical side to have a bend line since *only one* other *long side* of the base (4'–3') would have to be joined or welded.

Note: This is a *parallel-line development.*

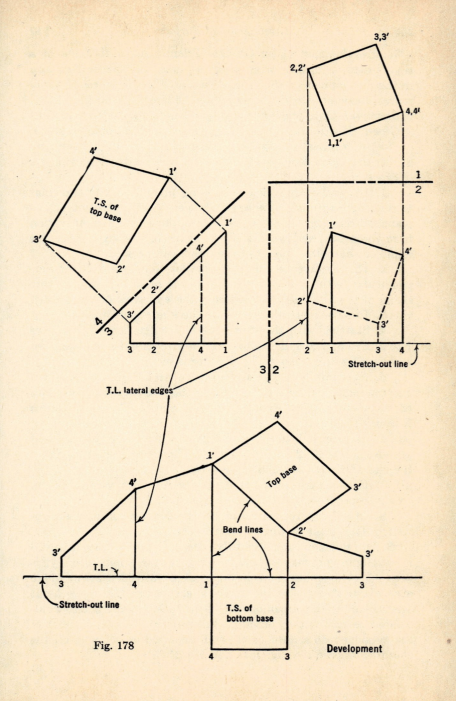

Fig. 178

Development

87. Development of an Oblique Prism. As in the case of a right prism, each face of an oblique prism will appear *true size* in the development. (See Fig. 179a and Fig. 179b for a pictorial representation of the development of the lateral surface.)

Example: Development of oblique prism. (Refer to Fig. 180.)

Given: Horizontal and vertical projections of an oblique prism.

Find: Develop the lateral surface of the oblique prism and attach the top and bottom bases to the surface development.

Procedure:

1. Draw the edge view of cutting plane S–S **perpendicular** to the *true length* lateral edges of the prism. The location of this plane is arbitrary (see view #2). This plane cuts a **right** section from the prism, and when the true shape of this right section is found the actual perimeter of the prism can be seen.

2. Draw reference line 2–3 parallel to plane S–S (or perpendicular to the T.L. lateral edges of the prism) to find the true shape of the plane section (actually the **end view** of the prism).

3. In a convenient space, draw the stretch-out line.

4. Starting with the shortest lateral edge, lay off distances 1 to 2, 2 to 3, 3 to 4, and 4 to 1 on the stretch-out line (measured in view #3 where the actual perimeter is seen).

5. At points 1, 2, 3, 4, and 1 on the stretch-out line (measuring from plane S–S in view #2) lay off the **true length** lateral edges of the prism: $1'$–$1''$, $2'$–$2''$, $3'$–$3''$, $4'$–$4''$, and $1'$–$1''$ (note distances X and Y).

6. Connect points $1'$, $2'$, $3'$, $4'$, and $1'$ with straight lines and likewise points $1''$, $2''$, $3''$, $4''$, and $1''$. This is the development of the lateral surface of the oblique prism.

7. Find the true shape of the top base by projection (bottom base appears true shape in the top view).

8. Attach the top and bottom bases to the surface development. This is the required development.

88. Development of a Right Cylinder. The principles involved in developing prisms can be applied to the development of cylinders. A cylinder can be considered as a multisided prism where each **element** of the cylinder is a lateral edge. The stretch-out line is equal to the circumference of the cylinder, and is either calculated or laid out by dividing the **right section** (end view) of the cylinder into a number of small arcs (actually laying off the chord distances; see Fig. 181a and Fig. 181b).

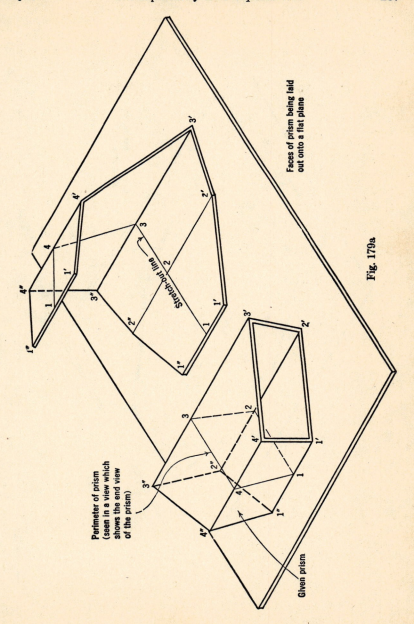

Faces of prism being laid
out onto a flat plane

Stretch-out line

Perimeter of prism
(seen in a view which
shows the end view
of the prism)

Given prism

Fig. 179a

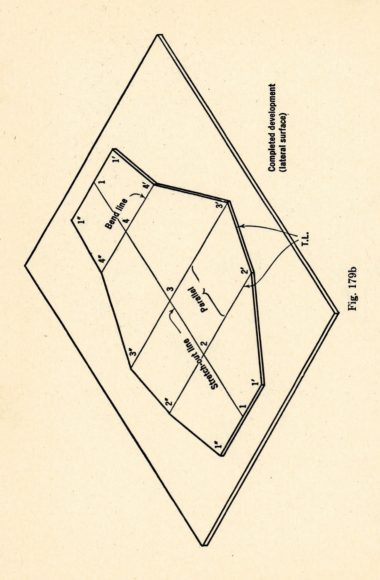

Completed development
(lateral surface)

Fig. 179b

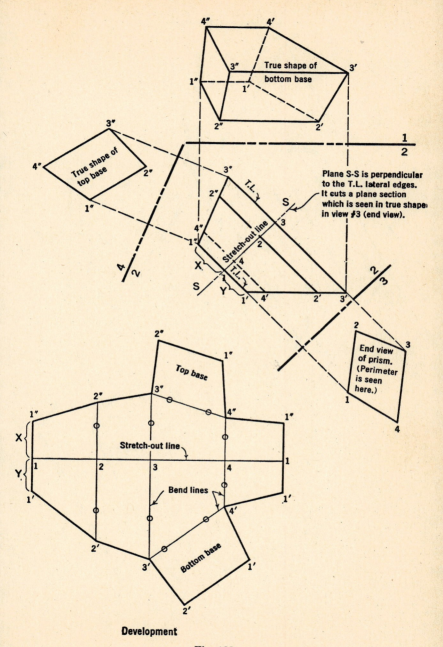

Development

Fig. 180

Within the figure:

4″ 4′

3″ True shape of
3′
bottom base
1″ 1′

2″ 2′

True shape of
top base
3″

4″ 2″

1″

Plane S-S is perpendicular
to the T.L. lateral edges.
It cuts a plane section
which is seen in true shape
in view #3 (end view).

T.L.
S
3″
2″
3
2
S
4″
1″
Stretch-out line
X
T.L.
4
2
Y
1′ 4′ 2′ 3′

2
3
End view
of prism.
(Perimeter
is seen
here.)
1
4

2″ 1″

Top base
3″ 4″

1″ 1″

X
Stretch-out line
1 2 3 4 1

Bend lines
Y
1′ 1′

2′ 4′

3′ Bottom base 1′

2′

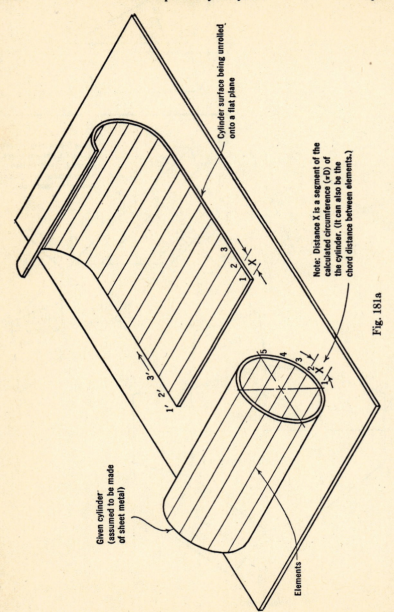

Cylinder surface being unrolled onto a flat plane

Note: Distance X is a segment of the calculated circumference (πD) of the cylinder. (It can also be the chord distance between elements.)

Given cylinder (assumed to be made of sheet metal)

Elements

Fig. 181a

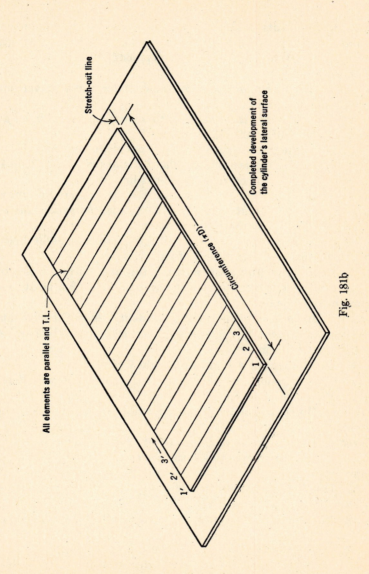

Fig. 181b

Example: Development of a right cylinder. (Refer to Fig. 182.)

Given: Horizontal and vertical projections of a right cylinder.

Find: Develop the lateral surface of the cylinder and attach the top and bottom bases to the development.

Procedure:

1. Divide the circumference of the cylinder into a convenient number of equal parts (see view #1).

2. View #2 shows the *true length elements* of the cylinder at each division (1–1′, 2–2′, 3–3′, etc.).

3. Draw the stretch-out line in a convenient location.

4. With dividers lay off the *chord distances* 1–2, 2–3, 3–4, 4–5, etc. (see view #1) on the stretch-out line.

5. At points 1, 2, . . ., 12 & 1 on the stretch-out line draw the *true length elements* 1–1′, 2–2′, 3–3′, etc., *perpendicular* to the stretch-out line at their proper points.

6. Connect points 1′, 2′, . . ., 12′ & 1′ with straight lines. This is the surface development of the cylinder.

7. Attach the *true shapes* of the top and bottom bases to the surface development (at arbitrary points). This is the required development.

Note: This is a *parallel-line development.* Since the elements of a cylinder are *parallel,* these elements are represented by parallel lines in the development.

89. Development of a Truncated Right Cylinder.

Example: Development of a truncated right cylinder. (Refer to Fig. 183 on p. 265.)

Given: Horizontal and vertical projections of a truncated right cylinder.

Find: Develop the lateral surface of the cylinder and attach the top and bottom bases to the surface development.

Procedure:

1. Divide the circumference of the cylinder into a convenient number of equal parts (see view #1).

2. View #2 shows the *true length elements* of the cylinder at each division (1–1′, 2–2′, 3–3′, etc.).

3. Draw the stretch-out line in a convenient location.

4. With dividers lay off the *chord distances* 1–2, 2–3, 3–4, 4–5, etc. (see view #1) on the stretch-out line.

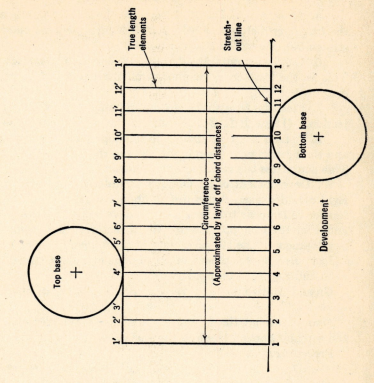

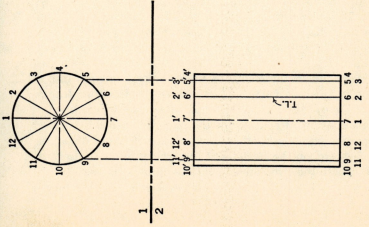

Fig 182

5. At points 1, 2, . . ., 12 & 1 on the stretch-out line draw the *true length elements* 1–1′, 2–2′, 3–3′, etc., *perpendicular* to the stretch-out line at their proper points.

6. Connect points 1′, 2′, . . ., 12′ & 1′ with a *smooth curve.* This is the surface development of the truncated right cylinder.

7. Find the true shape of the top base of the cylinder by projection. (Draw R.L. 2–3 parallel to the edge view of this base.)

8. Attach the true shapes of the top base (seen in view #3) and the bottom base (seen in view #1) to the surface development. This completes the required development.

90. Development of an Oblique Cylinder. As in the case of a right cylinder, each element of an oblique cylinder and its circumference will appear *true length* in the development. (See Fig. 184a and Fig. 184b on pp. 266–267 for a pictorial illustration of the development of the lateral surface of an oblique cylinder.)

Example: Development of an oblique cylinder. (Refer to Fig. 185a and Fig. 185b on pp. 268–269.)

Given: Horizontal and vertical projections of an oblique cylinder.

Find: Develop the lateral surface of the cylinder and attach the top and bottom bases to the surface development.

Procedure:

1. Draw the edge view of cutting plane *S–S perpendicular* to the cylinder axis. The location of this plane is arbitrary (see view #2). This plane cuts a right section from the cylinder, and when the true shape of the right section is seen the actual circumference of the cylinder is also seen (see view #1 — end view of the cylinder).

2. Divide the circumference of the cylinder (view #1) into a convenient number of parts. (View #2 shows the *true length* elements of the cylinder at each division: 1′–1″, 2′–2″, etc.)

3. In a convenient space draw the stretch-out line.

4. Starting with the shortest element, lay off chord distances 1–2, 2–3, 3–4, etc. on the stretch-out line (measured with dividers in view #1 where the circumference of the cylinder is seen).

5. At points 1, 2, . . ., 12 & 1 on the stretch-out line (meas-

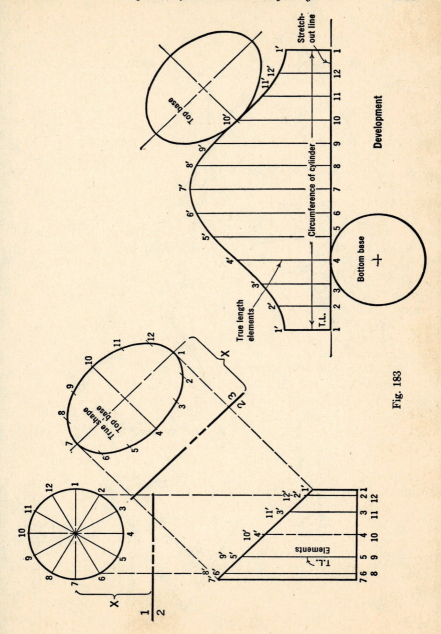

Fig. 183

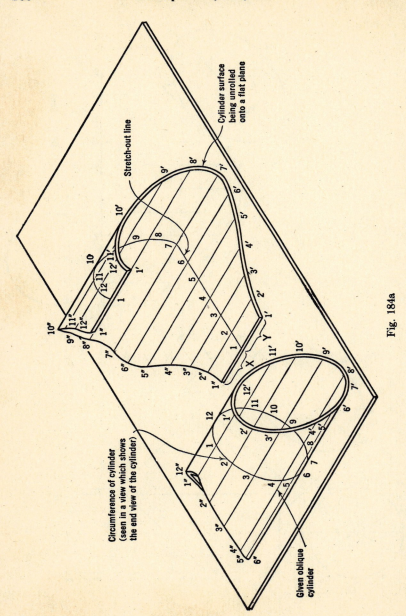

Cylinder surface being unrolled onto a flat plane

Stretch-out line

Circumference of cylinder (seen in a view which shows the end view of the cylinder)

Given oblique cylinder

Fig. 184a

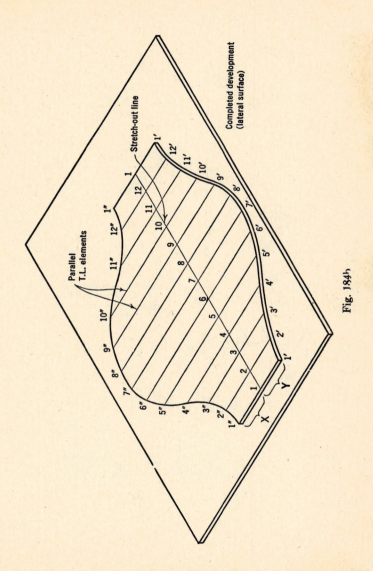

Fig. 184b

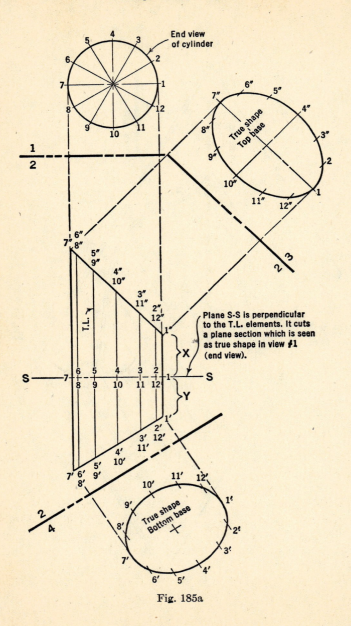

Fig. 185a

End view
of cylinder

True shape
Top base

Plane S-S is perpendicular
to the T.L. elements. It cuts
a plane section which is seen
as true shape in view #1
(end view).

T.L.

True shape
Bottom base

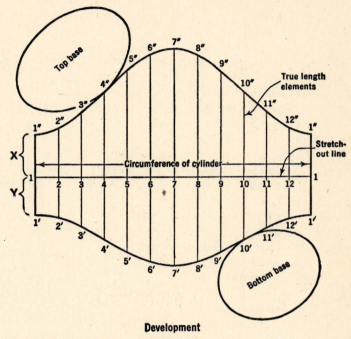

Development

Fig. 185b

uring from plane *S–S* in view #2) lay off the ***true length*** elements of the cylinder (1'–1'', 2'–2'', 3'–3'', etc.).　Note distances *X* and *Y*.

6. Connect points 1', 2', . . ., 12' & 1' with a smooth curve, and likewise points 1'', 2'', . . ., 12'' & 1''.　This is the development of the lateral surface of the cylinder.

7. Find the true shapes of the top and bottom bases by projection.

8. Attach the top and bottom bases to the surface development.　This is the required development.

91. Development of a Right Pyramid. The lateral surface of a pyramid consists of a number of *triangles* having a **common vertex**. To develop this lateral surface it is necessary to find the *true shapes* of these triangles. In a right pyramid all the *lateral edges* are of **equal length.** (See Fig. 186a and Fig. 186b for a pictorial representation of the development of the lateral surface of a right pyramid.)

Example: Development of a right pyramid. (Refer to Fig. 187.)

Given: Horizontal and vertical projections of a right pyramid.

Find: Develop the lateral surface of the pyramid and attach the base to the surface development.

Procedure:

1. Find the true length of one of the pyramid edges by the method of *revolution* (T.L. of edge *V*–3 found in example).

2. In a convenient space draw an arc having its center at *V* and a radius equal to the true length pyramid edge (in a right pyramid all the edges are of equal length).

3. With dividers lay off the *true length sides* of the base of the pyramid on the arc (distances 1 to 2, 2 to 3, 3 to 4, and 4 to 1, which are seen in true length in view #1).

4. Connect points 1, 2, 3, 4, and 1, and also points *V*–1, *V*–2, *V*–3, *V*–4, and *V*–1 with *straight lines*. This is the development of the lateral surface of the pyramid.

5. Attach the *true shape* of the pyramid base to the surface development. (Note that base is attached at one of its long sides.) This completes the required development.

Note: This is a *radial development.*

(Later in this chapter it will be shown how the method of radial development is applied to the development of a right cone. It is suggested that the student refer back to this section when he studies the development of a cone.)

92. Development of a Truncated Right Pyramid.

Example: Development of a truncated right pyramid. (Refer to Fig. 188.)

Given: Horizontal and vertical projections of a truncated right pyramid (rectangular base).

Find: Develop the lateral surface of the given pyramid and attach the top and bottom bases to the surface development.

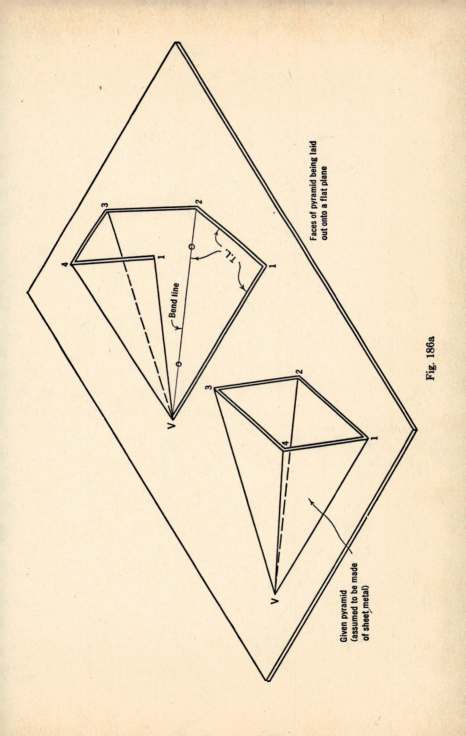

Faces of pyramid being laid
out onto a flat plane

Bend line

T.L.

Given pyramid
(assumed to be made
of sheet metal)

Fig. 186a

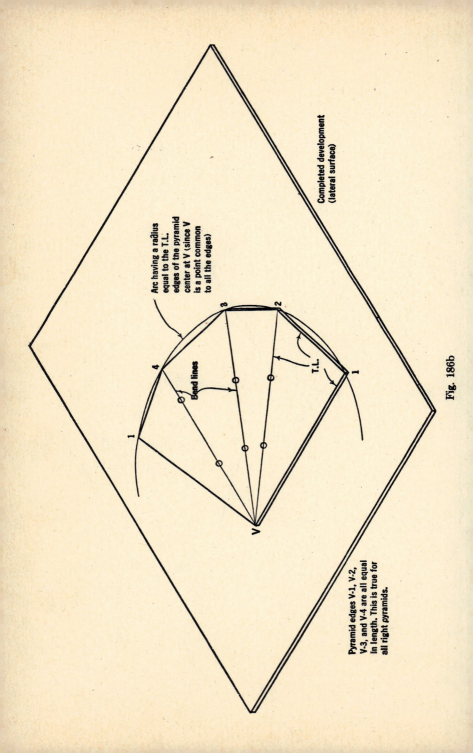

Completed development
(lateral surface)

Arc having a radius
equal to the T.L.
edges of the pyramid
center at V (since V
is a point common
to all the edges)

Bend lines

T.L.

Pyramid edges V-1, V-2,
V-3, and V-4 are all equal
in length. This is true for
all right pyramids.

Fig. 186b

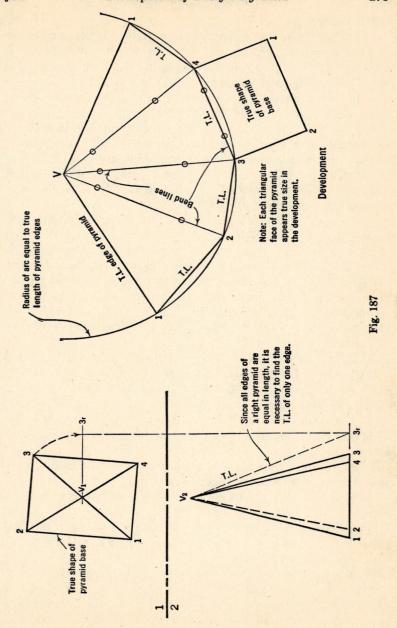

Radius of arc equal to true length of pyramid edges

T.L. edge of pyramid

Bend lines

Note: Each triangular face of the pyramid appears true size in the development.

True shape of pyramid base

Development

Fig. 187

True shape of pyramid base

Since all edges of a right pyramid are equal in length, it is necessary to find the T.L. of only one edge.

Procedure:

1. For purposes of construction, consider the given pyramid to be a **complete** right pyramid, having a vertex V. Find the **true length** of one of the pyramid edges by the method of revolution (T.L. of edge V–3 found in example).

2. In a convenient space draw an arc having a radius equal to the true length pyramid edge (V–3).

3. With dividers lay off on the arc the **true length sides** of the **bottom base** of the pyramid (distances 1 to 2, 2 to 3, 3 to 4, and 4 to 1 — true shape of bottom base seen in view #1).

4. Connect points 1, 2, 3, 4, and 1 with straight lines, and also connect V with points 1, 2, 3, 4, and 1. This is the development of the **complete** right pyramid.

5. By the method of revolution find distances v_2–$1_r'$, v_2–$2_r'$, v_2–$3_r'$, and v_2–$4_r'$ (view #2).

Note: Since the true length of edge v_2–3 has been found in view #2, and since all the lateral edges of a right pyramid are of equal length, any points on these edges can be projected (perpendicular to the axis of revolution) to a true length edge (v_2–3_r). Actual true length distances v_2–$1_r'$, v_2–$2_r'$, v_2–$3_r'$, and v_2–$4_r'$ can then be measured on the true length edge (see Chapter VI on Revolution).

6. On the surface of the **complete** right pyramid, lay off with dividers the **true length distances** V–$1'$, V–$2'$, V–$3'$, V–$4'$, and V–$1'$ on the proper T.L. edges (bend lines of the development of the complete right pyramid).

7. Connect points $1'$, $2'$, $3'$, $4'$, and $1'$ with **straight lines**. This is the development of the lateral surface of the given truncated right pyramid.

8. Find the **true shape** of the **top base** by projection (horizontal line seen as a point used in example).

9. Attach top base to the surface development at its longest side.

10. Attach the true shape of the bottom base (seen in view #1) to the surface development at one of its long sides. This completes the required development.

93. Development of an Oblique Pyramid (Truncated). As in the case of a right pyramid, each face of an oblique pyramid appears as **true shape** in the development. Since the lateral edges of an oblique pyramid are **not** all equal, it is necessary to find the

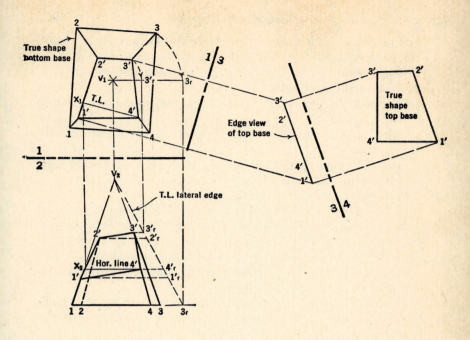

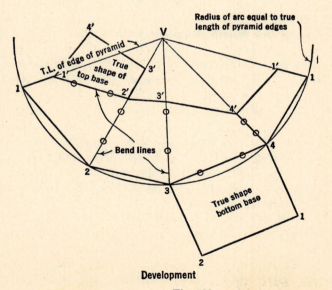

Development

Fig. 188

true length of *each* lateral edge before the surface development can be completed. (See Fig. 189a and Fig. 189b on pp. 277–278 for a pictorial representation of the surface development of a truncated oblique pyramid.)

Example: Development of a truncated oblique pyramid. (Refer to Fig. 190.)

Given: Horizontal and vertical projections of a truncated oblique pyramid.

Find: Develop the lateral surface of the given pyramid and attach the top and bottom bases to the surface development.

Procedure:

1. For purposes of construction, consider the given pyramid to be a *complete* oblique pyramid having a vertex V. Find the *true lengths* of *all* the lateral edges of the pyramid by the method of revolution (v_1 in view #1 used as a center of rotation).

2. Begin the surface development by drawing the *true length* of edge V–1 in a convenient space.

3. From point 1 (in the development) strike an arc equal to distance 1 to 2 (measured in view #1 where bottom base appears true shape).

4. Using V as a center, strike an arc equal to the true length of edge V–2. Where this arc intersects arc 1–2, is located point 2 in the development.

5. Using point 2 as a center, strike an arc equal to distance 2 to 3 (view #1).

6. Then using V as a center, strike an arc equal to the true length of edge V–3. Where this arc intersects arc 2–3, is located point 3 in the development.

7. Repeat the above procedure for each edge and side of the bottom base.

8. Connect points 1, 2, 3, 4, and 1 with straight lines, and also connect V with points 1, 2, 3, 4, and 1. This completes the surface development of the *complete* oblique pyramid.

9. Find the true length distances V–1′, V–2′, V–3′, and V–4′. (See note on page 274.)

10. With dividers, lay off distances V–1′, V–2′, V–3′, V–4′, and V–1′ on the proper true length lateral edges (bend lines) in the development.

11. Connect points 1′, 2′, 3′, 4′, and 1′ with straight lines. This completes the development of the *lateral* surface of the given truncated oblique pyramid.

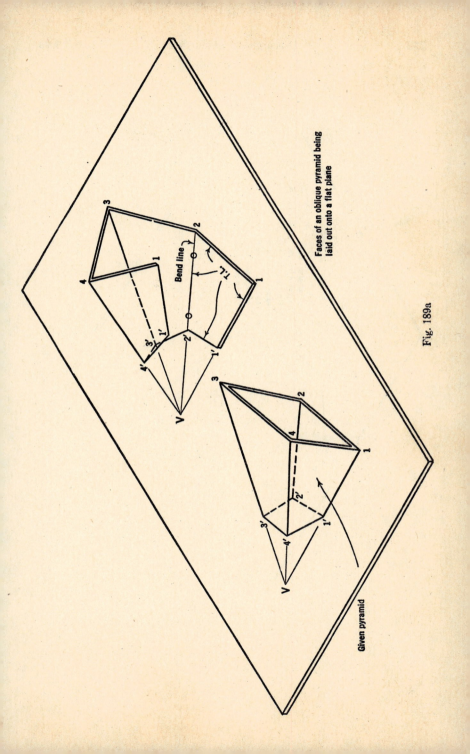

Faces of an oblique pyramid being laid out onto a flat plane

Given pyramid

Fig. 189a

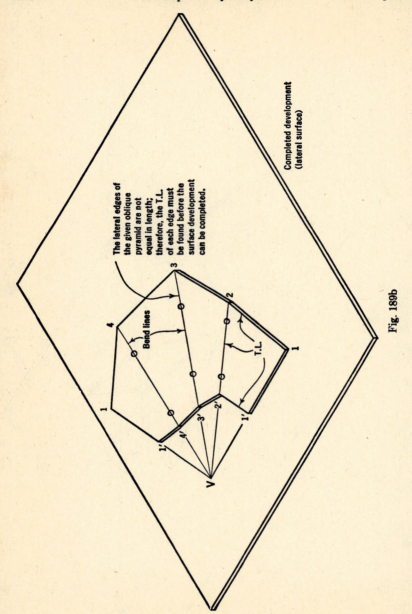

The lateral edges of the given oblique pyramid are not equal in length; therefore, the T.L. of each edge must be found before the surface development can be completed.

Bend lines

T.L.

Completed development (lateral surface)

Fig. 189b

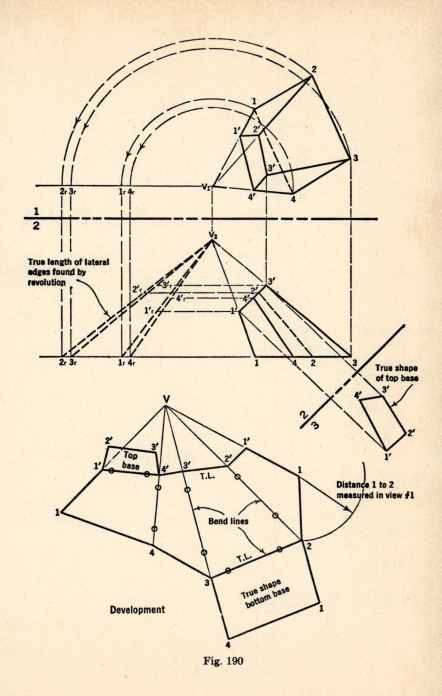

True length of lateral edges found by revolution

True shape of top base

True shape of bottom base

Distance 1 to 2 measured in view #1

Bend lines

Top base

Development

Fig. 190

12. Find the true shape of the top base by projection.

13. Attach the top base to the surface development at its longest side.

14. Attach the bottom base (true shape seen in view #1) to the surface development at its longest side. This completes the required development.

94. Development of a Right Cone. The principles which are used to develop the surface of a pyramid can be applied to develop the surface of a cone. A cone can be considered as a multisided pyramid; therefore, its lateral surface in effect is made up of an infinite number of *triangles*. In an actual development, the surface of the cone is divided into a convenient number of triangles and the *true shapes* of these triangles are found as they are in the development of a pyramid. The resulting development is an *approximation* since in actual practice the cone's surface can be divided into only a limited number of triangles. The more divisions that are made, the greater the accuracy of the development.

This method of dividing a surface into triangles is known as development by *triangulation*. (See Fig. 191a and Fig. 191b for a pictorial illustration of the development of the surface of a right cone.)

Example: Development of a right cone. (Refer to Fig. 192.)

Given: Horizontal and vertical projections of a right cone.

Find: Develop the surface of the cone and attach the base of the cone to the surface development.

Procedure:

1. Divide the base of the cone into a convenient number of equal parts (see view #1).

2. Draw elements v_1–1, v_1–2, v_1–3, etc.

3. In a convenient space, draw an arc having a radius equal to the *true length* of one of the *elements* of the cone. (Element V–4 appears T.L. in view #2.)

4. With dividers, lay off the *chord distances* 1 to 2, 2 to 3, 3 to 4, etc. (measured in view #1) on the arc just drawn.

5. Connect V–1, V–2, . . ., V–12 & V–1 with straight lines (representing the elements of the cone). This is the development of the surface of the cone.

6. Attach the *true shape* of the base of the cone (seen in view #1) to the surface development. This completes the required development.

Note: This is a *radial development*. Compare this development

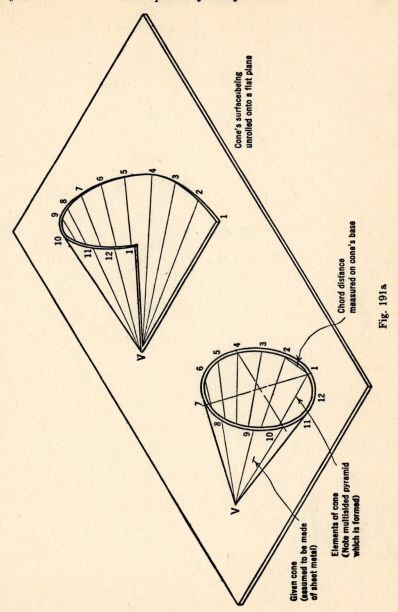

Cone's surface being
unrolled onto a flat plane

Chord distance
measured on cone's base

Given cone
(assumed to be made
of sheet metal)

Elements of cone
(Note multisided pyramid
which is formed)

Fig. 191a

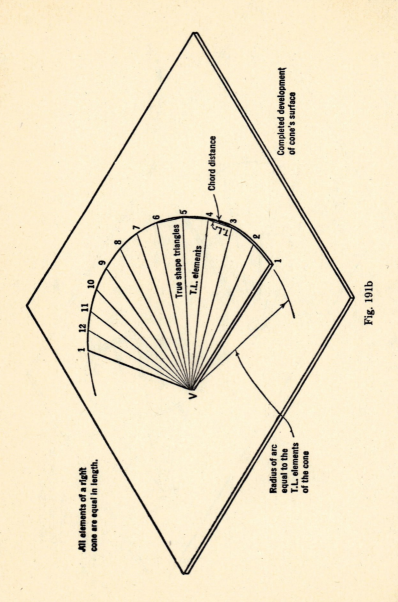

All elements of a right cone are equal in length.

Chord distance

True shape triangles

T.L. elements

Radius of arc equal to the T.L. elements of the cone

Completed development of cone's surface

Fig. 191b

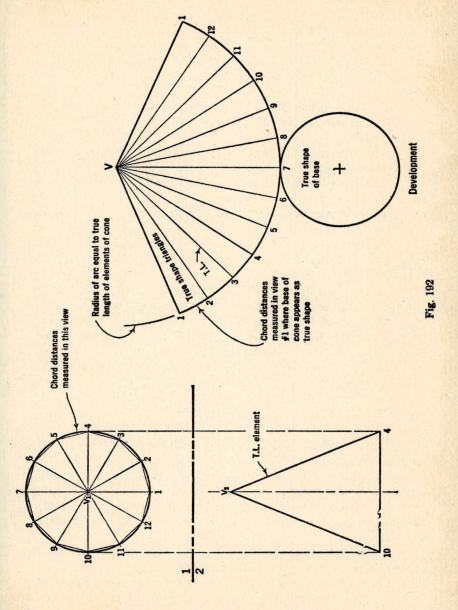

Fig. 192

with the development of a right pyramid (see Fig. 186a and b and Fig. 187, pp. 271–273).

95. Development of a Truncated Right Cone.

Example: Development of a truncated right cone. (Refer to Fig. 193 on p. 285.)

Given: Horizontal and vertical projections of a truncated right cone.

Find: Develop the surface of the given cone and attach the top and bottom bases of the cone to the surface development.

Procedure:

1. For purposes of construction, consider the given cone to be a *complete* right cone, having a vertex V.

2. Divide the base of the cone into a convenient number of equal parts (see view #1).

3. Draw elements V–1, V–2, V–3, etc. (see views #1 and #2).

4. In a convenient space draw an arc having a radius equal to a *true length element* of the cone (element V–1 appears T.L. in view #2).

5. With dividers, lay off the *chord distances* 1 to 2, 2 to 3, 3 to 4, etc. (measured in view #1) on the arc just drawn.

6. Connect V–1, V–2, V–3, etc., with straight lines (representing the T.L. elements of the complete right cone). This is the actual development of the *complete* right cone.

7. By the method of revolution find distances V–1′, V–2′, V–3′, etc. (see views #1 and #2).

8. On the surface development of the *complete* cone, lay off with dividers the *true length distances* V–1′, V–2′, V–3′, etc., on the proper elements.

9. Connect points 1′, 2′, 3′, etc., with a *smooth curve*. This is the development of the surface of the given truncated cone.

10. Find the *true shape* of the top base by projection and attach this base to the surface development.

11. Attach the *true shape* of the bottom base (seen in view #1) to the surface development. This completes the required development.

96. Development of an Oblique Cone (Truncated).

As in the case of a right cone, the surface of an oblique cone is divided into a number of triangles and the *true shapes* of these triangles are found (similar to the development of a *multisided pyramid*).

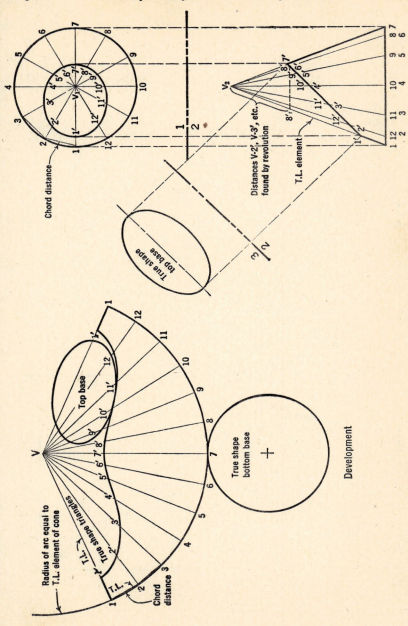

Distances V-2', V-3', etc., found by revolution

T.L. element

True shape top base

Chord distance

True shape bottom base

Radius of arc equal to T.L. element of cone

True shape triangles

T.L.

Chord distance

Top base

Development

Fig. 193

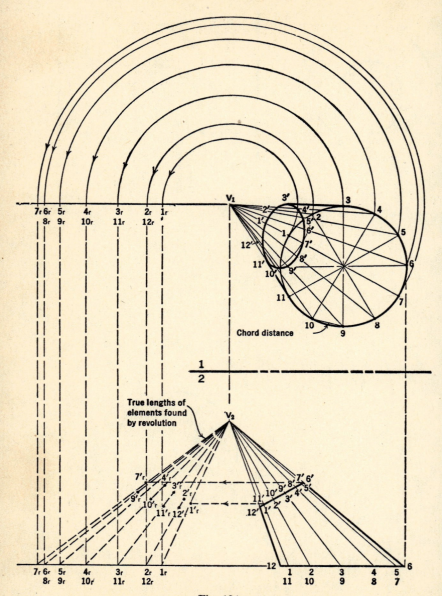

Chord distance

True lengths of elements found by revolution

Fig. 194a

Since the elements of an oblique cone are *not* all equal in length, it is necessary to find the true lengths of each element before the surface development can be completed.

Example: Development of a truncated oblique cone. (Refer to Fig. 194a and Fig. 194b.)

Given: Horizontal and vertical projections of a truncated oblique cone.

Find: Develop the surface of the given cone.

Procedure:

1. For purposes of construction, consider the given cone to be a *complete* oblique cone having a vertex V. Divide the base of the cone into a convenient number of parts (see view #1).

2. Draw elements V–1, V–2, V–3, etc. (see views #1 and #2).

3. Find the *true lengths* of *all* the elements by the method of revolution (v_1 in view #1 used as a center of rotation).

4. Begin the development by drawing the *true length* element V–1 (shortest element) in a convenient space.

5. From point 1 (in the development) strike an arc equal to distance 1 to 2 (measured in view #1 where bottom base appears true shape).

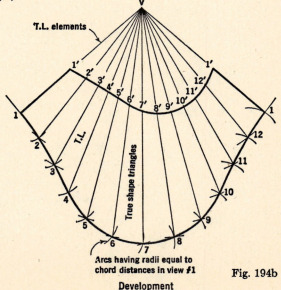

Fig. 194b

Development

6. Using V as a center, strike an arc equal to the true length of element V–2. Where this arc intersects arc 1–2, is located point 2 in the development.

7. Using point 2 (in the development) strike an arc equal to chord distance 2 to 3 (view #1).

8. Using V as a center, strike an arc equal to the true length of element V–3. Where this arc intersects arc 2–3, is located point 3 in the development.

9. Repeat the above procedure for each element and chord distance.

10. Connect points 1, 2, . . ., 12 & 1 with a smooth curve. This completes the surface development of the *complete* oblique cone.

11. Find the true length distances V–1′, V–2′, V–3′, etc. by the method of revolution (see views #1 and #2).

12. With dividers, lay off distances V–1′, V–2′, V–3′, etc. on the proper true length elements in the development.

13. Connect points 1′, 2′, . . ., 12′ & 1′ with a smooth curve. This completes the required surface development.

97. Development of a Transition Piece * (**Having a Warped Surface**). The surface of a warped transition piece cannot be accurately developed. However, it is possible to approximately

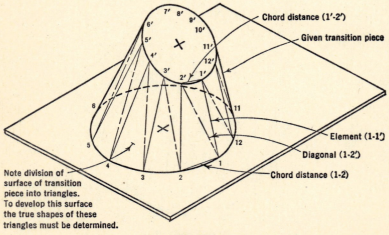

Fig. 195

* Usually a sheet metal duct which makes a transition from one type of opening to another.

develop such a surface by the method of *triangulation*. The surface of the transition piece is divided into a number of *triangles* and the *true shapes* of these triangles are found. The procedure is basically similar to that used in the development of pyramids and cones. (See Fig. 195.)

Example: Development of a transition piece having a warped surface. (Refer to Figs. 196, 196a, 196b, 196c, and 196d.)

Given: Horizontal and vertical projections of a transition piece.

Find: Develop the surface of the transition piece.

Procedure:

See
Fig. 196

1. Find the *true shape* of the top base of the transition piece by projection.

2. Divide the *true shape* of the top base and the *true shape* of the bottom base (seen in view #1) into a convenient number of equal parts. Number each division (refer to views #1 and #2).

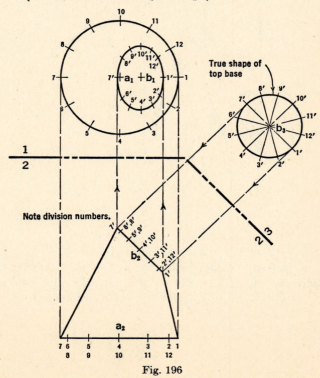

Fig. 196

See
Fig. 196a

3. Connect *like-numbered* divisions with *straight lines* (1–1', 2–2', 3–3', etc.). These lines can be considered as *elements* on the surface of the transition piece.

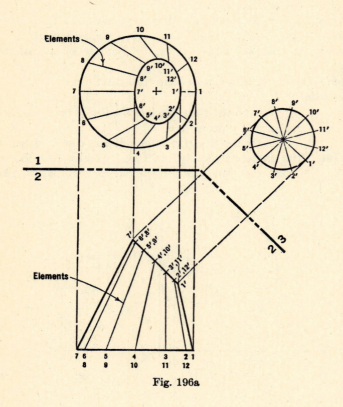

Fig. 196a

See
Fig. 196b

4. Draw *diagonal* lines (1–2', 2–3', 3–4', etc.) between each pair of elements in views #1 and #2. It is here that the surface of the transition piece is actually divided into triangles. (Note chord distances.)

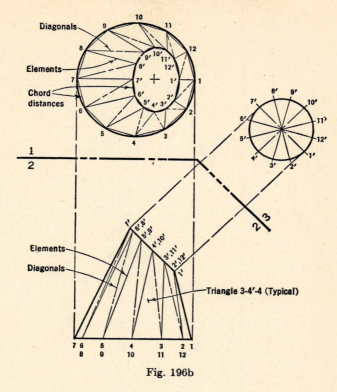

Fig. 196b

5. Find the *true lengths* of each *element* and *diagonal* by the method of revolution.

Note: The axis of revolution is drawn in a space away from the given views to eliminate the overlapping of lines. (See Fig. 196c.) Distance X is the *top view length* of element 5–5'. It is only necessary to measure distance X with dividers and to transfer this distance to the new location of the axis of revolution as shown in Fig. 196c. This procedure for finding the true lengths of lines is a short-cut method of revolution which eliminates the necessity of drawing the actual arcs of rotation.

See
Fig. 196c

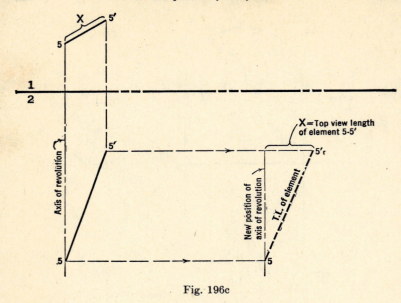

Fig. 196c

See Development Fig. 196d {

6. Draw an arc having a radius equal to the *true length* of *element* 1–1′. (Represent 1–1′ with a straight line.)

7. From point 1′, strike an arc having a radius equal to the *true length chord distance* 1′–2′ (measured in view #3).

8. From point 1, strike an arc having a radius equal to the *true length diagonal* 1–2′. The point at which this arc intersects arc 1′–2′ is the location of point 2′.

9. From point 2′, strike an arc having a radius equal to the *true length element* 2–2′.

10. From point 1, strike an arc having a radius equal to the *true length chord distance* 1–2 (measured in view #1). The point at which this arc intersects arc 2–2′ is the location of point 2.

11. Continue the above procedure for all *elements, diagonals,* and *chord distances.* (Note that what is being done is actually the laying out

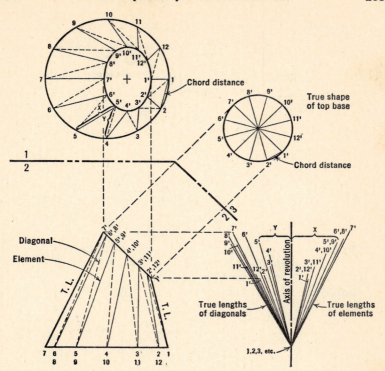

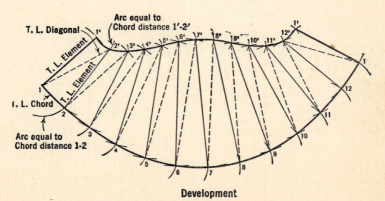

Development

Fig. 196d

of the *true shapes* of the individual *triangles* into
which the given surface was divided.)

See
Development
Fig. 196d

12. Connect points 1, 2, . . ., 12 & 1 with
a smooth curve, and likewise points 1', 2', . . .,
12' & 1'. This completes the required development
of the given transition piece.

Practice Problems

X-1. Given: Horizontal and vertical projections of a truncated right
prism having a hexagonal bottom base.

  Find: (a) Develop the lateral surface of the given prism.

  (b) Attach the top and bottom bases to the surface
development.

Suggestion: Use two sheets of paper for this solution.

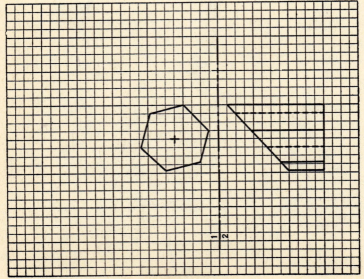

X-1

X-2. Given: Horizontal and vertical projections of an oblique prism.
Find: (a) Develop the lateral surface of the given prism.
(b) Attach the top and bottom bases to the surface development.
Suggestion: Use two sheets of paper for this solution.

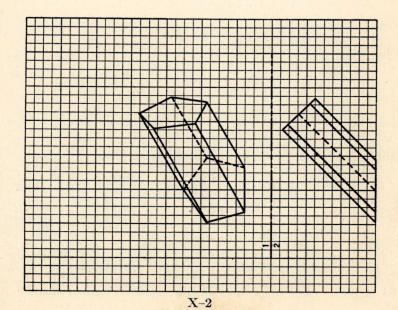

X-2

X–3. Given: Horizontal and vertical projections of a truncated right cylinder (axis AB).

      Find: (a) Develop the surface of the given cylinder.

          (b) Attach top and bottom bases to the surface development.

      Suggestion: Use two sheets of paper for this solution.

X–4. Given: Horizontal and vertical projections of an oblique cylinder (axis LM).

      Find: (a) Develop the surface of the given cylinder.

          (b) Attach the top and bottom bases to the surface development.

      Suggestion: Use two sheets of paper for this solution.

X–5. Given: Horizontal and vertical projections of a truncated right pyramid (vertex V).

      Find: (a) Develop the lateral surface of the given pyramid.

          (b) Attach top and bottom bases to the surface development.

      Suggestion: Use two sheets of paper for this solution.

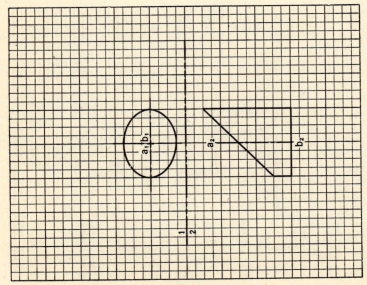

X–3

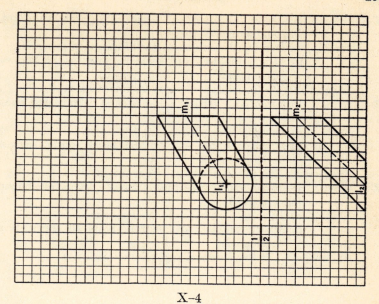

X–4

X–5

X–6. Given: Horizontal and vertical projections of a truncated oblique pyramid (vertex *O*).

    Find: (a) Develop the lateral surface of the given pyramid.

          (b) Attach the top and bottom bases to the surface development.

Suggestion: Use two sheets of paper for this solution.

X–7. Given: Horizontal and vertical projections of a truncated right cone (vertex *V*).

    Find: (a) Develop the surface of the given cone.

          (b) Attach the top and bottom bases to the surface development.

Suggestion: Use two sheets of paper for this solution.

X–8. Given: Horizontal and vertical projections of a truncated oblique cone (vertex *O*).

    Find: (a) Develop the surface of the given cone.

          (b) Attach the top and bottom bases to the surface development.

Suggestion: Use two sheets of paper for this solution.

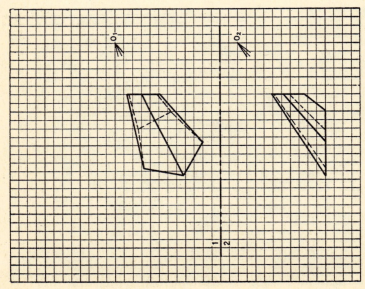

X–6

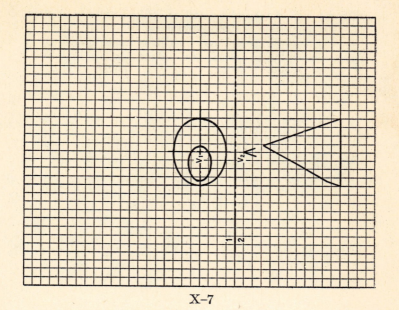

X–7

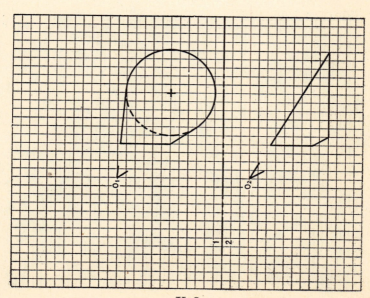

X–8

X-9. Given: Horizontal and vertical projections and a partial profile projection of a transition piece.

Find: Develop the surface of the given transition piece.

Suggestion: Use two sheets of paper for this solution.

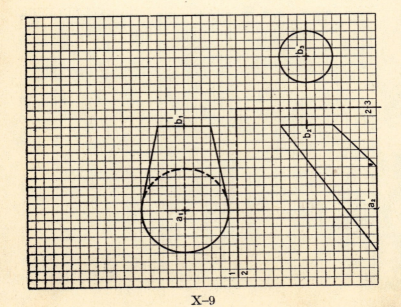

X–9

X–10. Given: Horizontal and vertical projections of a transition piece (circular and rectangular openings).

Find: Develop the lateral surface of the given transition piece.

Suggestions: Use two sheets of paper for this solution. Draw elements from points *ABCD* to the circular opening and find true lengths of these elements. Develop surface by means of triangulation.

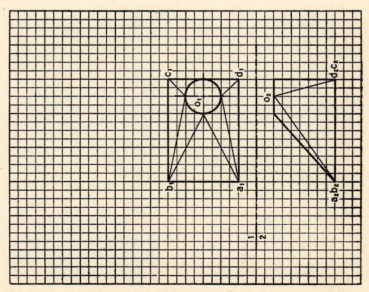

X–10

Appendix

Included in this appendix are: an example of an analytical method for finding the shortest distance between two points, a number of examples showing how the principles of descriptive geometry can be applied to perspective drawing and the projection of shadows, and applications dealing with various fields of engineering.

No attempt has been made to elaborate on the different subjects presented in this appendix. The material here is intended to be introductory in nature. The engineering application problems have been set up so that the student can solve them and check his solution with the solution presented at the end of each example.

I. ANALYTICAL METHOD (SHORTEST DISTANCE BETWEEN TWO POINTS, OR THE TRUE LENGTH OF A LINE)

Since descriptive geometry deals with space problems, its relation to solid (or space) analytic geometry is apparent. Descriptive geometry, as we have seen, is the *graphical* approach to the solution of space problems, whereas analytic geometry (dealing with space) is an *analytical* approach to the same space problems.

As in the case of descriptive geometry, *mutually perpendicular planes* are used as the basis of space analytic geometry. These planes are known as *coordinate planes,* and the lines of intersection of these planes are called the *axes of coordinates* (axes X, Y, and Z). The point (O) at which axes X, Y, and Z intersect is called the *origin.*

The location of a point in space, for example, can be fixed relative to the coordinate planes. All measurements which are necessary to locate a point are taken from the origin, along the X, Y, and Z axes, and *perpendicular* to the coordinate planes. The *perpendicular distances* from the point in space to the coordinate planes are the *coordinates* of the point in space.

Fig. 197 shows the location of a point P in space having coordinates $x = 2$, $y = 2$, and $z = 3$. This is normally noted as $P(2,2,3)$.

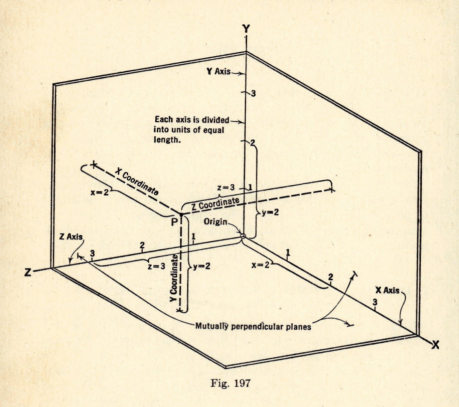

Fig. 197

Example: Shortest distance between two points — analytical method. (Refer to Fig. 198 and Fig. 199.)

Given: Two points L and M having the following coordinates:

$$L \text{ coordinates} = (2,1,3).$$
$$M \text{ coordinates} = (1,3,1).$$

Find: Determine analytically the shortest distance between points L and M.

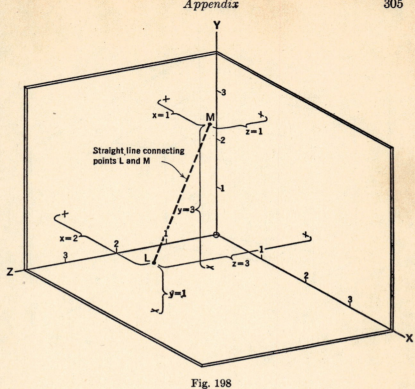

Fig. 198

Procedure:

    1. Locate points L and M relative to the coordinate planes.

    2. Draw a straight line connecting points L and M.

    3. Construct a prism whose faces are *parallel* to the *coordinate planes* and having line LM as a *diagonal* (see Fig. 199).

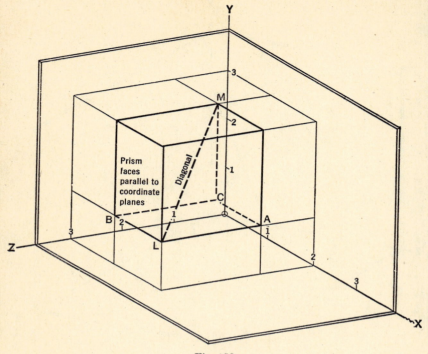

Fig. 199

4. Determine the dimensions of the prism by the coordinates which fix the length of each side of the prism. For example:

$$LA = (z = 3) - (z = 1) = (z = 2).$$

$$\therefore LA = 2 \text{ units.}$$

See
Fig. 200

$$LB = (x = 2) - (x = 1) = (x = 1).$$
$$\therefore LB = 1 \text{ unit.}$$

$$MC = (y = 3) - (y = 1) = (y = 2).$$
$$\therefore MC = 2 \text{ units.}$$

5. Solve for diagonal length LM:

(a) $(LC)^2 = (LA)^2 + (LB)^2.$
(b) $(LM)^2 = (LC)^2 + (MC)^2.$

Substituting $(LC)^2 = (LA)^2 + (LB)^2$ in equation (b) we get:

$$(LM)^2 = (LA)^2 + (LB)^2 + (MC)^2.$$
$$LM = \sqrt{(LA)^2 + (LB)^2 + (MC)^2}.$$

See Fig. 200

6. Substituting values of LA, LB, and MC found in step #4:

$$LM = \sqrt{(2)^2 + (1)^2 + (2)^2} = \sqrt{9}.$$
$$LM = 3 \text{ units.}$$

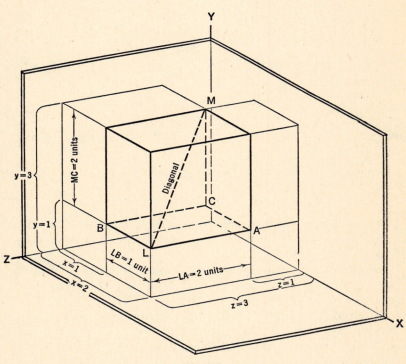

Fig. 200

II. PERSPECTIVE DRAWING

In orthographic projection the lines of sight of an observer are assumed to be *parallel,* meaning that the position of the observer, relative to the object he is looking at, is at *infinity.* (See Fig. 201.)

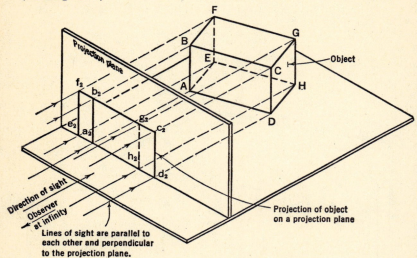

Fig. 201

In actuality, when an observer looks at an object, his lines of sight are *not parallel* but *radiate* from the point of observation (his eye). (See Fig. 202.)

If a drawing is made reproducing what the observer *actually sees,* the result will be a picture similar to a photograph. This picture will be a *perspective drawing.*

A projection plane. called *picture plane,* is placed between the observer and the given object. The point of observation is called the *station point.* Where each line of sight *pierces* the picture plane, a point is located on the perspective drawing. (See Fig. 203 for a pictorial illustration.)

In constructing a perspective drawing the *horizon* is considered to be at *eye level* (the distance of the station point above the ground plane). Fig. 204 illustrates this and also shows how the *ground line* appears (line of intersection of the picture plane and the ground plane upon which the object rests).

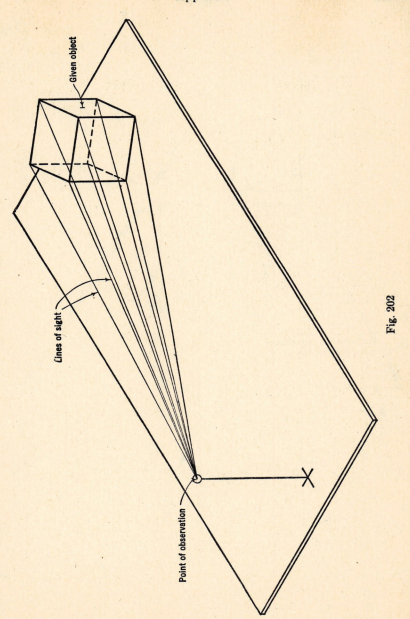

Given object

Lines of sight

Point of observation

Fig. 202

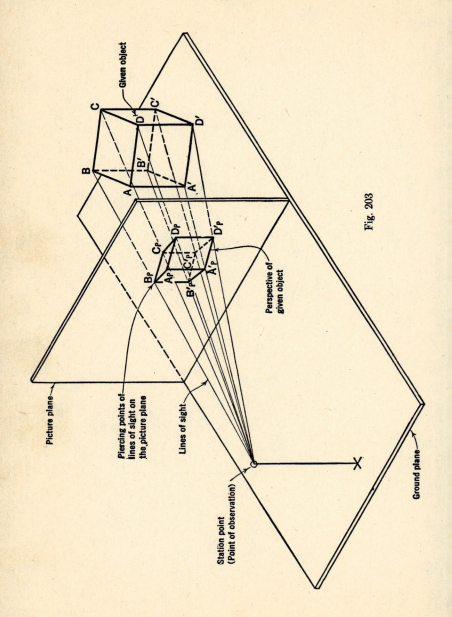

Given object

Picture plane

Piercing points of
lines of sight on
the picture plane

Lines of sight

Station point
(Point of observation)

Perspective of
given object

Ground plane

Fig. 203

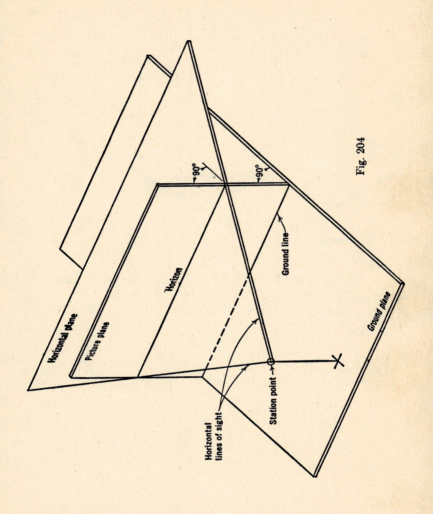

Fig. 204

Basic Types of Perspective Drawing.

A. Parallel Perspective. One face of an observed object is parallel to the picture plane (see Fig. 205).

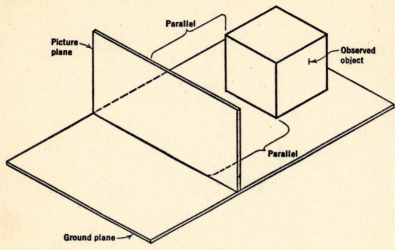

Fig. 205

B. Angular Perspective. The vertical edges of an observed object are parallel to the picture plane, but the rest of the faces of the object are *not* parallel to the picture plane (see Fig. 206).

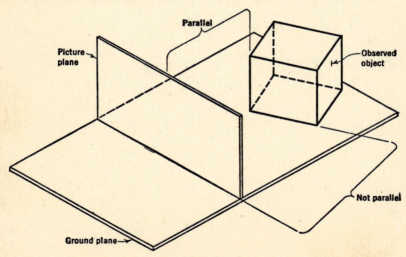

Fig. 206

Another form of angular perspective results when no face or vertical edge of an observed object is parallel to the picture plane (see Fig. 207).

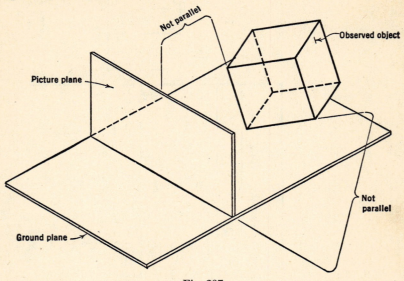

Fig. 207

Angular Perspective Drawing. The most common type of perspective drawing is the angular perspective where only the vertical edges of an object are parallel to the picture plane. This type of perspective is shown in the example below. The method of direct projection is used to construct the perspective drawing in the example.

Example: Angular perspective of a cube (direct projection method). (Refer to Fig. 208.)

Given: Horizontal and profile projections of a cube and station point *SP*.

Find: Draw the perspective of the given cube.

Note: The position of the station point is given in this example. When the student is left to choose the station point he must be careful to make a good choice, since the position from which the observer views an object will determine exactly how much and how well he will see the object. This is analogous to photography, where the camera angle is so important in taking a good picture.

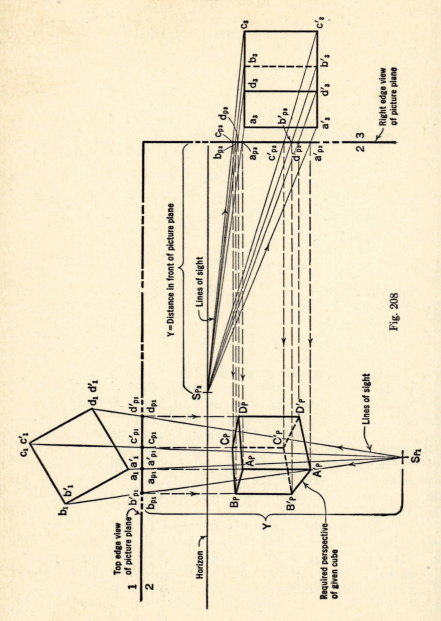

Fig. 208

Procedure:

1. Draw lines of sight from sp_1 to points a_1, b_1, c_1, and d_1 on top view of given cube.

2. Note where each line of sight pierces the top edge view of the picture plane (at a_{p1}, b_{p1}, c_{p1}, and d_{p1}).

3. Draw lines of sight from sp_3 to points a_3, b_3, c_3, and d_3 on profile view of given cube.

4. Note where each line of sight pierces the profile edge view of the picture plane.

5. To locate A_p (in perspective) draw projectors from a_{p1} and a_{p3}. The point at which these projectors intersect is A_p. Repeat this procedure for points B_p, C_p, and D_p.

6. Connect points A_p, B_p, C_p, and D_p with straight lines. This is the top surface of the cube in perspective.

7. Repeat the above procedure for points a'_{p1}, b'_{p1}, c'_{p1}, d'_{p1}, and a'_{p3}, b'_{p3}, c'_{p3}, and d'_{p3}.

8. Connect all points with straight lines. This is the required perspective of the given cube.

Vanishing Points. Another method of constructing a perspective drawing involves the use of *vanishing points*. A vanishing point may be defined as the point at which a *horizontal line* (on an observed object), when extended, disappears at the *horizon*. Figs. 209 and 209a illustrate pictorially how vanishing points may be determined.

Example: Angular perspective of a cube (vanishing point method). (Refer to Fig. 210.)

Given: Horizontal and profile projections of a cube, the horizontal projection of station point *SP*, and the horizon line.

Find: Draw the perspective of the cube using the vanishing point method.

Procedure: (Note that cube is touching picture plane.)

1. Draw a horizontal line of sight from sp_1 parallel to a_1b_1 and piercing the top edge view of the picture plane at x_{p1}.

2. From x_{p1} draw a line perpendicular to the horizon line. The foot of the perpendicular is VP_L. This is the *left vanishing point.*

3. Draw a horizontal line of sight from sp_1 parallel to a_1d_1 and piercing the top edge view of the picture plane at y_{p1}.

4. From y_{p1} draw a line perpendicular to the horizon line.

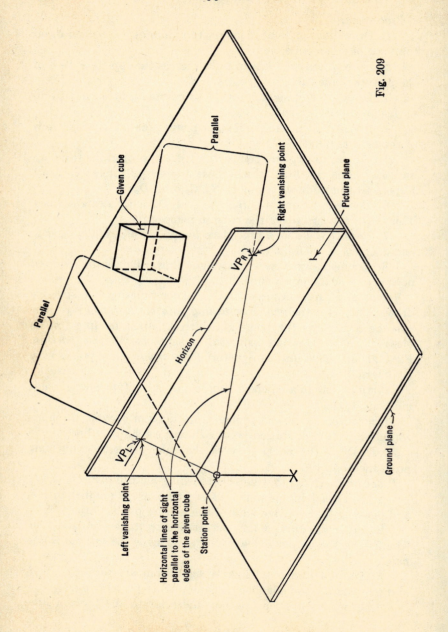

Given cube

Parallel

Parallel

Right vanishing point

Picture plane

Horizon

VP_R

VP_L

Left vanishing point

Horizontal lines of sight parallel to the horizontal edges of the given cube

Station point

Ground plane

X

Fig. 209

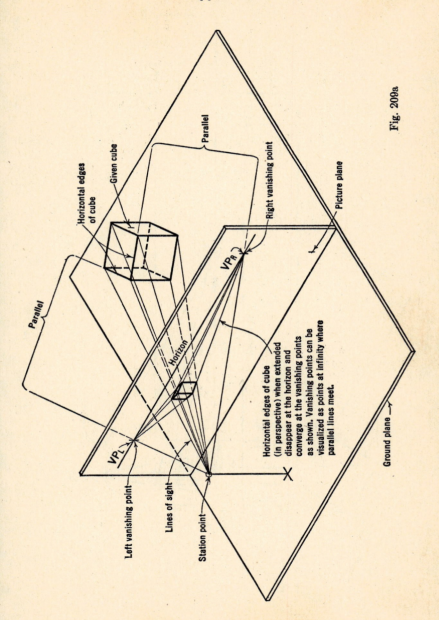

Given cube

Horizontal edges of cube

Parallel

Parallel

Right vanishing point

VP$_R$

Picture plane

Horizon

Horizontal edges of cube (in perspective) when extended disappear at the horizon and converge at the vanishing points as shown. Vanishing points can be visualized as points at infinity where parallel lines meet.

VP$_L$

Left vanishing point

Lines of sight

Station point

Ground plane

Fig. 209a

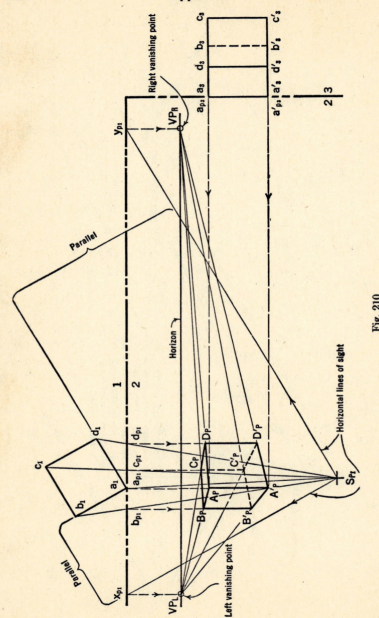

Fig. 210

The foot of the perpendicular is VP_R. This is the **right vanishing point**.

5. From sp_1 draw lines of sight to a_1, b_1, c_1, and d_1. (Note piercing points a_{p1}, b_{p1}, c_{p1}, and d_{p1} on picture plane.)

6. To locate A_p draw projectors from a_{p1} and a_{p3}. The point at which these projectors intersect is A_p. Repeat this procedure for point A'_p. Connect A_p and A'_p with a straight line.

7. From A_p and A'_p draw straight lines to the left and right vanishing points. (These are horizontal lines which disappear at the vanishing points.)

8. To locate points B_p and B'_p draw projectors from b_{p1}. The points at which these projectors intersect the vanishing horizontal lines are B_p and B'_p. (D_p and D'_p are located in like manner.)

9. From B_p and B'_p draw straight lines to the right vanishing point. From D_p and D'_p draw straight lines to the left vanishing point. The points at which these two sets of lines intersect locate C_p and C'_p respectively.

10. Connect all points with straight lines. This is the required perspective of the given cube.

Note: By the use of vanishing points the actual construction of the perspective is simplified, since the positions of the horizontal lines in the perspective drawing are determined by the vanishing points. Since two vanishing points were used in the example, this method is also known as two-point perspective.

III. SIMPLE SHADES AND SHADOWS

In the following presentation on shades and shadows it is assumed that light rays are parallel and travel in straight lines (source of light assumed to be at infinity).

When light rays are intercepted by an object, a section of the object (on the "light" side) will be illuminated, while another section of the object (opposite the "light" side) will be shielded from the light rays. The shielded section is called *shade,* and the points at which lines representing the intercepted light rays pierce any plane beyond the object determine the *cast shadow* of the object.

Definitions Pertaining to Shades and Shadows. The following definitions illustrate the concepts of shade and shadow. (Refer to Fig. 211.)

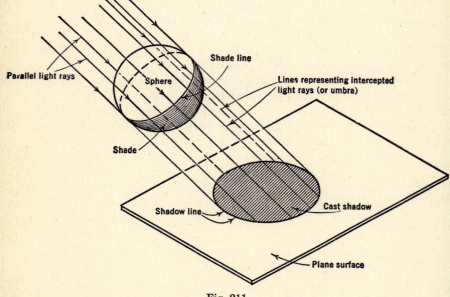

Fig. 211

Shade: The part of the surface of an object from which light is excluded by the object itself.

Umbra: The *space* behind an object from which light is excluded.

Cast Shadow: The intersection of the umbra with **any other surface.** (Descriptive geometry principle involved: **Piercing point of a line and a plane.**)

Cast Shadow of a Point on a Plane Surface. The point of intersection of a line representing a light ray (which is intercepted by a given point) and a plane surface is the cast shadow of the given point. (Refer to Fig. 212.)

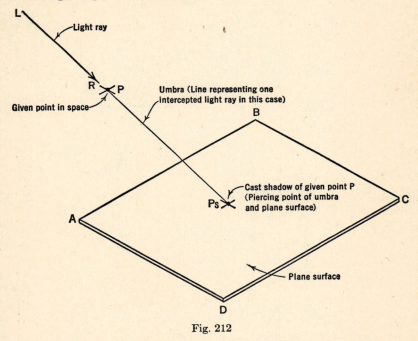

Fig. 212

Example: Cast shadow of a point on a plane surface. (Refer to Fig. 213.)

Given: Horizontal and vertical projections of point *P*, light ray *LR*, and plane surface *ABCD*.

Find: Cast shadow of point *P* on plane *ABCD*.

Procedure:

Apply the principle of finding the piercing point of a line and a plane. (Vertical cutting plane method is used in example.)

Cast Shadow of a Point on a Cylinder. The point of intersection of a line representing a light ray (which is intercepted by a given point) and a cylinder is the cast shadow of the point on the cylinder. (See Fig. 214.)

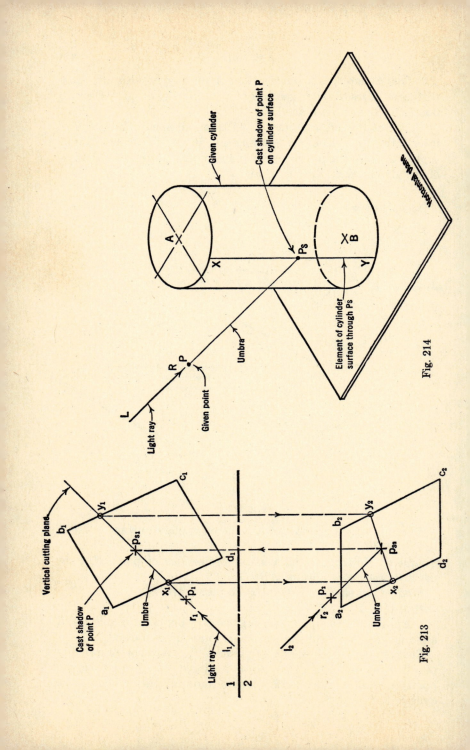

Fig. 214

Fig. 213

Example: Cast shadow of a point on a cylinder. (Refer to Fig. 215.)

Given: Horizontal and vertical projections of point P, cylinder AB, and light ray LR.

Find: Cast shadow of point P on cylinder AB.

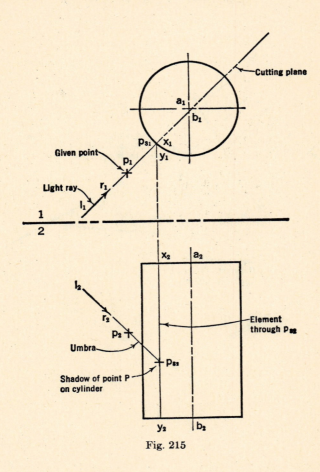

Fig. 215

Procedure:

Apply the principle of finding the piercing point of a line and a cylinder. (Cutting plane method is used in example.)

Cast Shadow of a Line on a Plane Surface. The light rays that are intercepted by a line lie in one plane; therefore the umbra also lies in one plane. The line of intersection between the umbra and a given plane surface is the cast shadow of the given line on the plane surface. (See Fig. 216.)

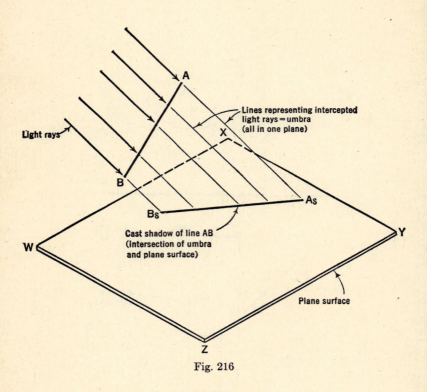

Fig. 216

Example: Cast shadow of a line on a plane surface. (Refer to Fig. 217.)

Given: Horizontal and vertical projections of line *AB*, plane *WXYZ*, and direction of light rays.

Find: Cast shadow of line *AB* on plane *WXYZ*.

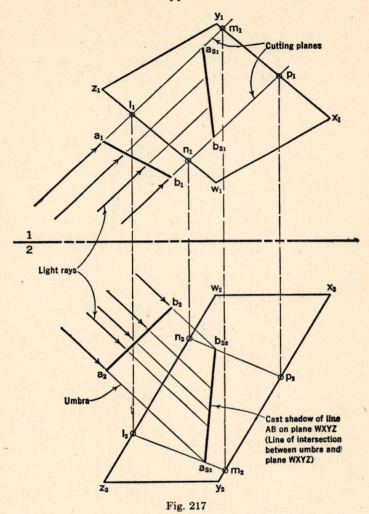

Fig. 217

Procedure:

Find where the lines representing intercepted light **rays** (umbra) intersect plane *WXYZ*. (Vertical cutting plane **method** is used in example.)

Cast Shadow of a Line on a Cylinder. The light rays which are intercepted by a given line lie in one plane, and therefore the umbra also lies in the same plane. The line of intersection between the umbra and a given cylinder is the cast shadow of the line on the cylinder. (See Fig. 218.)

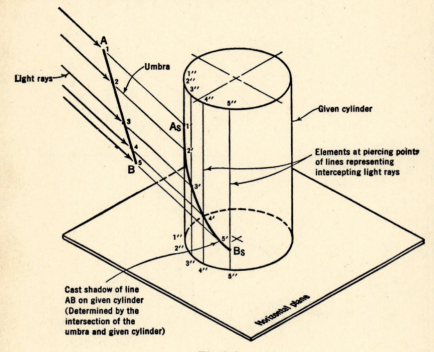

Fig. 218

Example: Cast shadow of a line on a cylinder. (Refer to Fig. 219.)

Given: Horizontal and vertical projections of line AB, a right cylinder, and direction of light rays.

Find: Shadow of line AB on given cylinder.

Procedure:

1. Draw a convenient number of light rays intercepted by line AB. Locate these rays in both views by projection (1, 2, 3, 4, and 5).

2. Find where these rays pierce the given cylinder in view #1 (at points $1'$, $2'$, $3'$, $4'$, and $5'$).

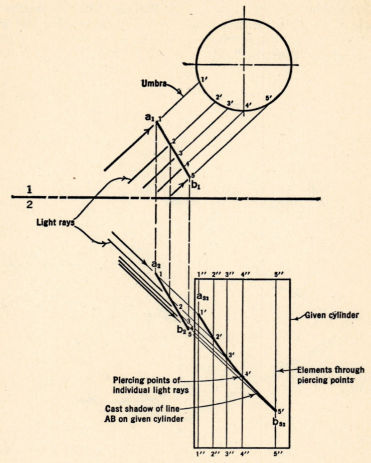

Fig. 219

3. Determine points of intersection in view #2. (Draw elements 1″–1″, 2″–2″, 3″–3″, 4″–4″, and 5″–5″. Where ray #1 intersects element 1″–1″ is a point on the cast shadow of line *AB*. Repeat this procedure for all rays and elements.)

4. Connect points 1′, 2′, 3′, 4′, and 5′ in view #2 with a smooth curve. This is the required cast shadow of line *AB* on the given cylinder.

Note: The principle involved in the solution of this example is the piercing point of a line and a cylinder.

Cast Shadow of a Plane on a Plane Surface. The cast shadow of a plane on a plane surface is determined by the piercing points of the lines representing intercepted light rays (umbra) on the plane surface. (See Fig. 220.)

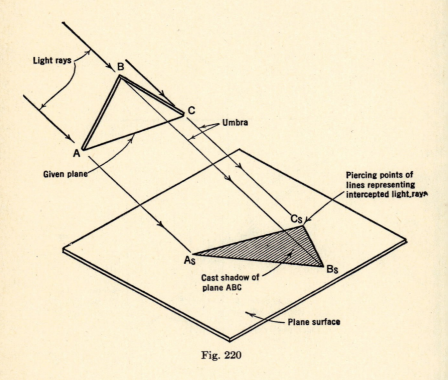

Fig. 220

Example: Cast shadow of a plane on a plane surface. (Refer to Fig. 221.)

Given: Horizontal and vertical projections of plane *ABC*, plane surface *WXYZ*, and direction of light rays.

Find: Cast shadow of plane *ABC* on plane *WXYZ*.

Procedure:

Find the piercing points of the lines representing light rays intercepted by points *A*, *B*, *C* of the given plane on plane *WXYZ* by the cutting plane method.

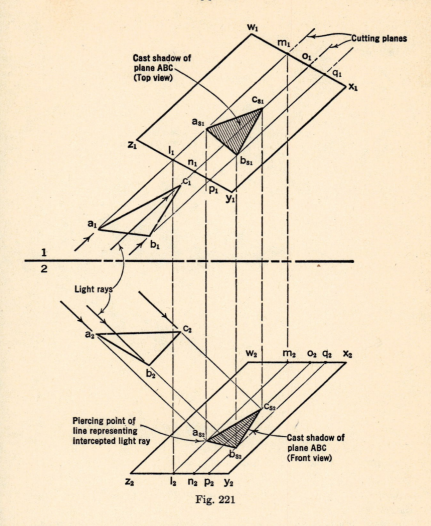

Fig. 221

Shade and Shadow of a Solid Having Plane Surfaces. The *shade line* is determined by the *edges* of the solid and usually can be determined by inspection. The shadow is determined by finding the piercing points of the lines representing intercepted light rays with the plane on which the shadow is cast. (See Fig. 222.)

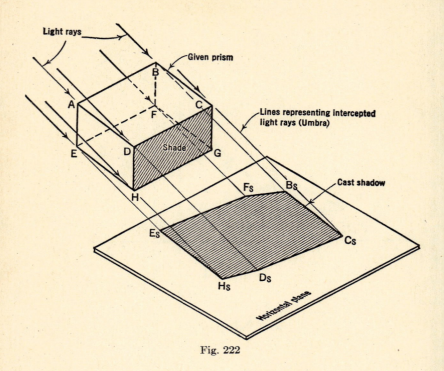

Fig. 222

Example: Shade and shadow of a solid having plane surfaces. (Refer to Fig. 223.)

Given: Horizontal and vertical projections of a prism and direction of light rays.

Find: Shade of prism and cast shadow of prism on a horizontal plane.

Procedure:

1. Draw light rays to points *A, B, C, D* and *E, F, G, H* in views #1 and #2.

2. Determine shade by inspection. (Interception of light rays by line *HD* excludes light from face *CDGH*.)

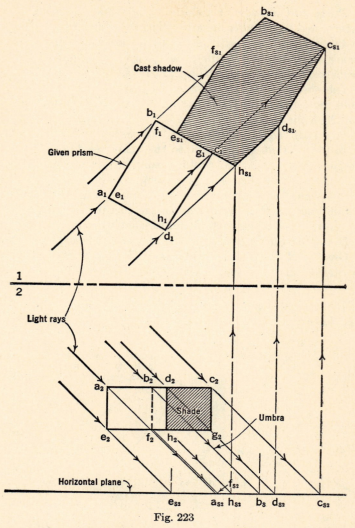

Fig. 223

3. Find piercing points (E_S, F_S, B_S, C_S, D_S, and H_S) of lines representing intercepted light rays (umbra) on given horizontal plane by projection. (Points A_S and G_S fall within the shadow outline.)

4. Connect above points with straight lines. These form the shadow line which determines the required cast shadow.

Shade and Shadow of a Cone. The shade lines of a cone are determined by parallel light rays that are tangent to the surface of the cone. The shade lines actually are two elements on the cone's surface at the points of tangency of the light rays. These elements intercept the parallel light rays.

The shadow line is determined by the piercing points of the lines representing the above intercepted light rays and the plane on which the shadow is cast.

The lines representing these intercepted light rays actually form planes that are tangent to the cone (see Fig. 224). These tangent planes in actual problem solution are used to determine two elements at their points of tangency (the shade lines) and the shadow line, which is the line of intersection of the tangent planes and the plane on which the shadow is cast. (See Sec. 69 for principles dealing with planes tangent to cones.)

Example: Shade and shadow of a cone. (Refer to Fig. 225.)

Given: Horizontal and vertical projections of a right cone (vertex *V*) and direction of light rays.

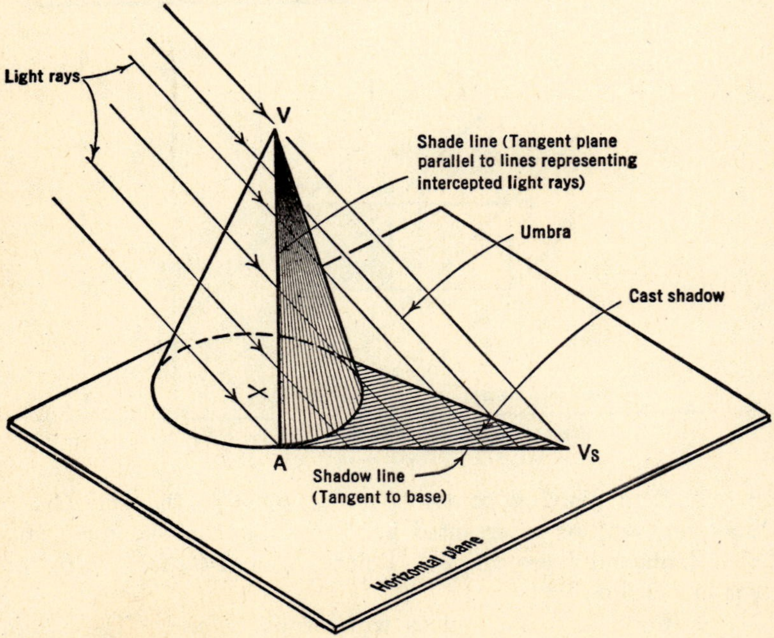

Light rays

V

Shade line (Tangent plane parallel to lines representing intercepted light rays)

Umbra

Cast shadow

Vs

A

Shadow line (Tangent to base)

Horizontal plane

Fig. 224

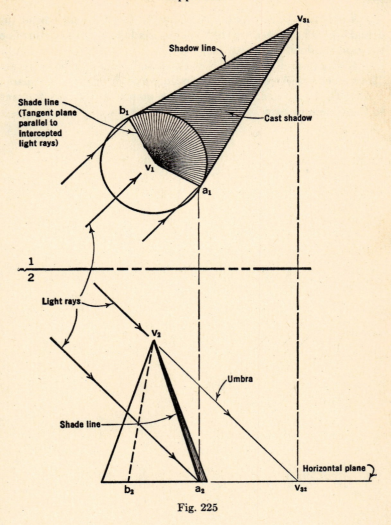

Fig. 225

Find: Shade and shadow of given cone on a horizontal plane.
Procedure:

1. Draw a light ray through the vertex in view #2.

2. Find where the line representing this light ray (intercepted by v_2) pierces the given horizontal plane in views #1 and #2 by projection (at v_{S1} and v_{S2}).

3. From v_{s1} draw two lines tangent to the cone's base (at a_1 and b_1). These lines are the required shadow lines and determine the cast shadow.

4. Connect VA and VB with straight lines in both views. These lines are the required shade lines and determine the shade. VA, VB, V_SA, and V_SB determine planes which are tangent to the cone and parallel to the given light rays.

IV. MISCELLANEOUS ENGINEERING APPLICATIONS

The following examples will serve to illustrate further how the principles of descriptive geometry can be applied to solve various engineering problems. If care is exercised in the construction of the graphical solution, the results, from a practical point of view, will compare in accuracy to the results that can be obtained analytically.

The problems which follow should be set up orthographically and solved by the student. At the end of each problem setup is a completed solution which can serve as a check for the student.

Guy Wire Problem (pictorially illustrated in Fig. 226). AB is a lead-in wire attached to a horizontal antenna at point A. Point D is an anchor for a guy wire attached to the top of the antenna mast at point E. The minimum clearance allowable between the lead-in and guy wires is 2′-6″. (See Fig. 228 for completed solution.)

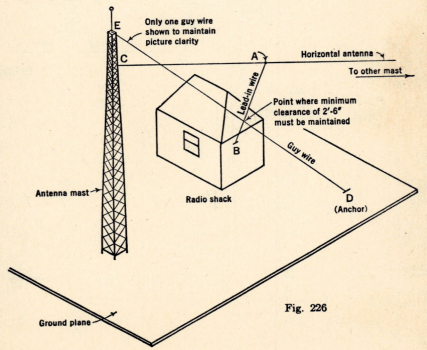

Fig. 226

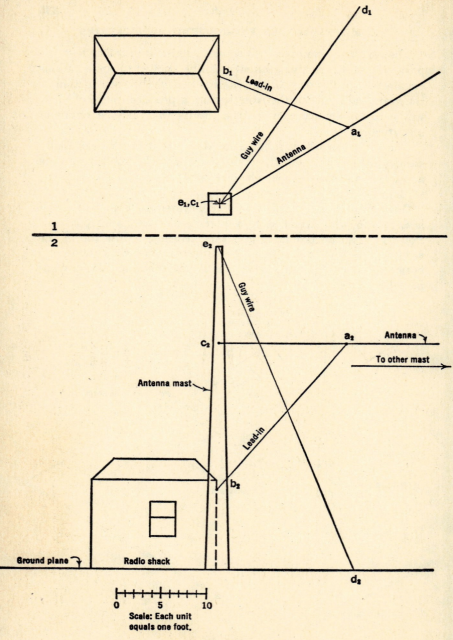

Fig. 227

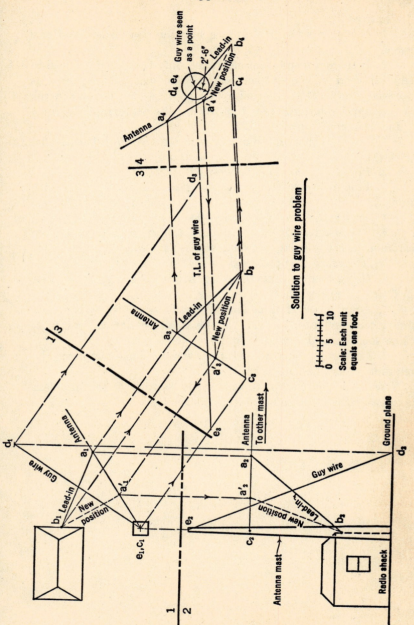

Scale: Each unit equals one foot.

Solution to guy wire problem

Fig. 228. See procedure on p. 338.

Required:

1. Find the present clearance between the lead-in and guy wires.

2. If there is insufficient clearance, only the end of the lead-in (*A*) attached to the antenna can be moved. If movement is necessary, show the new position of point *A* in all views in the orthographic solution. (See Fig. 227 for the orthographic setup.)

Procedure: Refer to Fig. 228.

1. Find the true length of guy wire DE (pass RL 1–3 parallel to d_1e_1).

2. Project a_1b_1 and a_1c_1 to view #3.

3. Find guy wire DE as a point (pass RL 3–4 perpendicular to d_3e_3).

4. Project a_3b_3 and a_3c_3 to view #4.

5. The shortest distance from a_4b_4 to d_4e_4 is measured on the common perpendicular between these two wires. (Note that existing clearance is approximately zero.)

6. Using d_4e_4 as a center draw a circle having a radius of 2′–6″ (required clearance).

7. Draw a line from b_4 tangent to above circle. Where this line intersects antenna (a_4c_4) is the new location of point *A*. (New location is designated as a'_4.)

8. Project a'_4 back to all other views. This completes the requirements of the problem.

Note: The principle involved in the solution of this problem is the shortest distance between two skew lines.

Pulley Bracket Problem (general orientation pictorially illustrated in Fig. 229). *AB* and *BC* are known cable directions in one section of an airplane and *WXYZ* is a face of a frame member in the airplane. The problem is to design a bracket which can be mounted on face *WXYZ* to support a 4″ diameter pulley which will guide the cable in the given directions.

The following data must be obtained:

1. The angle the pulley makes with the face *WXYZ*.

2. The location of the center of the pulley. (See Fig. 230 for the orthographic setup and Fig. 231 for the completed solution.)

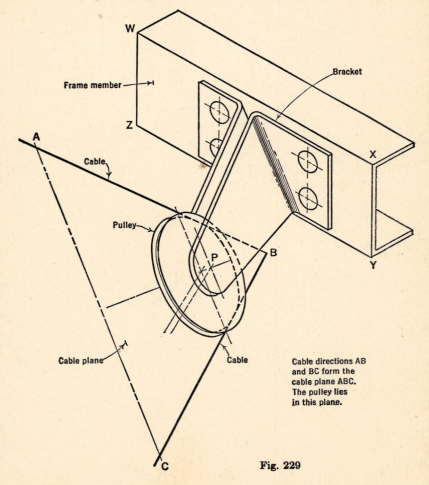

Cable directions AB
and BC form the
cable plane ABC.
The pulley lies
in this plane.

Fig. 229

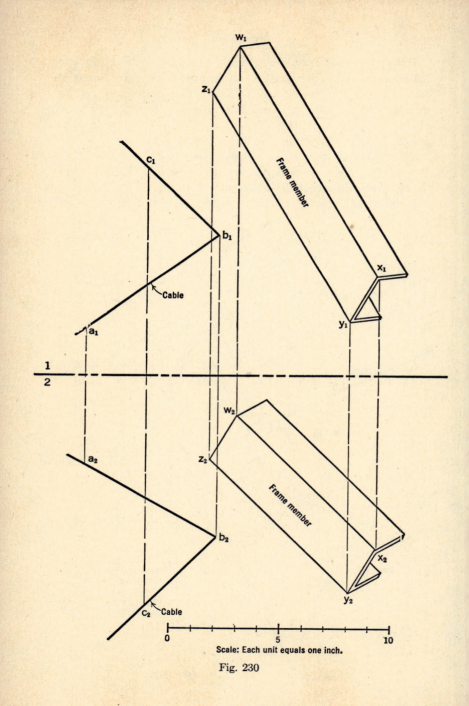

Fig. 230

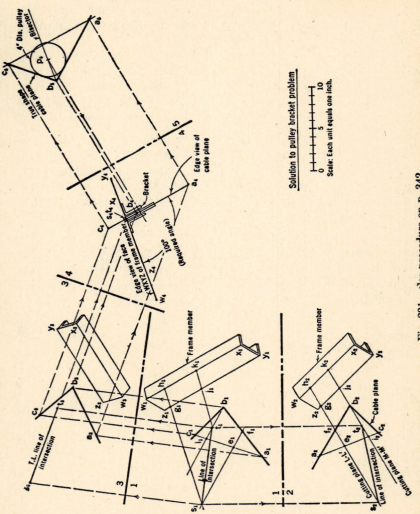

Solution to pulley bracket problem

Scale: Each unit equals one inch.

Fig. 231. See procedure on p. 342.

Procedure: Refer to Fig. 231.

Angle between two planes

1. Determine the line of intersection (ST) of the cable plane ABC and the plane face $WXYZ$. (Consider both planes to be indefinite. Draw cutting planes $L-L'$ and $M-M'$ through ABC and $WXYZ$.)

2. Draw the edge views of the cable plane ABC and face $WXYZ$. [Find the true length (s_3t_3) of the line of intersection and then see it as a point (s_4t_4).] In view #4 the angle between ABC and $WXYZ$ can be measured.

Location of pulley center

3. Draw the true shape of the cable plane ABC (RL 4–5 parallel to $a_4b_4c_4$).

4. In the true shape view (#5) bisect the angle formed by a_5b_5 and b_5c_5. Locate the center (p_5) of the 4″ diameter pulley on this bisector.

5. Project p_5 and the pulley back to view #4.

All required data have been found, and the pulley bracket now can be designed to fulfill the given conditions. It is suggested that the student complete this problem by designing and completely dimensioning the required pulley bracket.

Ventilator Duct Problem (pictorially illustrated in Fig. 232). *A* and *B* are the centers of two circular openings in a floor spaced 4'-0" apart. These openings are to accommodate two ventilator ducts, having outside diameters of 12", whose outlets are flush with the top of the floor. The two ducts are to be attached to a single blower duct, which is located 4'-6" below the top of the floor in line with centers *A* and *B*. The connection is to be made with a 90 degree *Y*-branch having 12" diameter legs. The floor is 3" thick, and the 2" × 10" floor joists are spaced 16" on center. (Ducts are centered between the joists.)

Design a *Y*-branch made of sheet metal which will satisfy the above conditions and also allow a 1" clearance between the *Y*-branch legs and the floor joists. (See Fig. 234 for completed solution.)

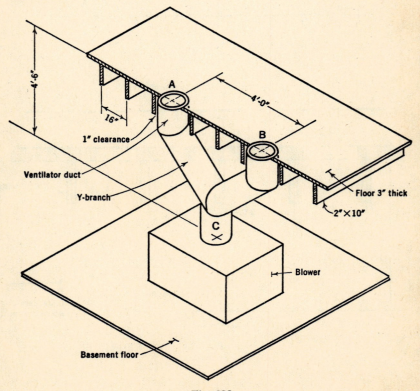

Fig. 232

Required data for design purposes:

1. Minimum lengths of ventilator ducts to allow 1″ of clearance between the Y-branch legs and the floor joists.

2. Lines of intersection between the ducts and the Y-branch components.

3. Development of the surface of the ducts and the Y-branch components so that templates can be made for the sheet metal man. (See Fig. 233 for the orthographic setup and Fig. 234 for completed solution.)

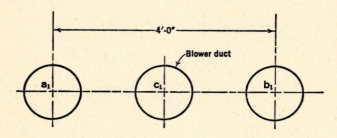

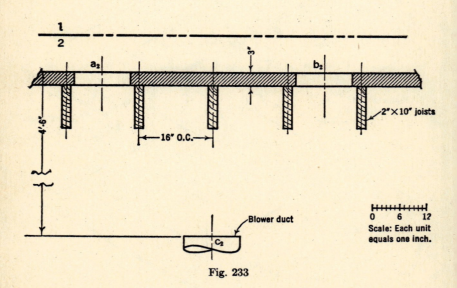

Fig. 233

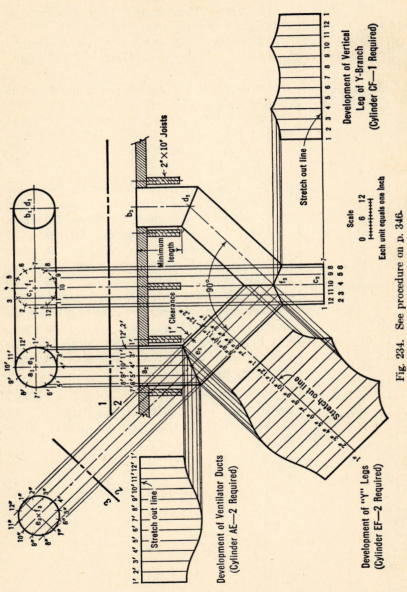

Fig. 234. See procedure on p. 346.

Procedure: Refer to Fig. 234.

Clearance

> 1. Draw 12″ diameter circles at centers *A*, *B*, and *C*, and complete the top view of the ventilator ducts and blower duct (see view #1).
> 2. Project 12″ circles (top view of ducts) to view #2.
> 3. Swing 1″ arcs from corners of floor joists to fix the required clearance (see view #2).
> 4. Draw 45 degree lines tangent to 1″ arcs (view #2).
> 5. Complete front view of ducts and *Y*-branch (view #2).

Lines of intersection

> 6. Lines of intersection between the ventilator ducts and the *Y*-branch components can be determined by inspection. Since the ducts and the *Y*-branch components are parts of cylinders of like diameters, the lines of intersection are plane curves (which are ellipses in this case, seen as edges in view #2). When the true lengths of the axes of the cylinders are seen (as they are in view #2) the lines of intersection appear as straight lines. (The student can prove this by passing cutting planes parallel to the cylinder axes to determine the lines of intersection by projection.)

Development

> 7. The ventilator ducts and the *Y*-branch components, being parts of cylinders, may all be developed by the methods illustrated in Chapter X.

Index